I0110599

The EXXON VALDEZ Oil Spill

A Report to the President

from
Samuel K. Skinner
Secretary, Department of Transportation
and
William K. Reilly
Administrator, Environmental Protection Agency

May 1989

Prepared By
The National Response Team

The Exxon Valdez Oil Spill

A Report to the President

from

Samuel K. Skinner
Secretary, Department of Transportation

and

William K. Reilly
Administrator, Environmental Protection Agency

Prepared by

The National Response Team

May 1989

CONTENTS

APPENDICES

JMKYVIW

EXECUTIVE SUMMARY

Shortly after midnight on March 24, 1989, the 987-foot tank vessel Exxon Valdez struck Bligh Reef in Prince William Sound, Alaska. What followed was the largest oil spill in U.S. history. The oil slick has spread over 3,000 square miles and onto over 350 miles of beaches in Prince William Sound, one of the most pristine and magnificent natural areas in the country. Experts still are assessing the environmental and economic implications of the incident. The job of cleaning up the spill is under way, and although the initial response proceeded slowly, major steps have been taken.

The very large spill size, the remote location, and the character of the oil all tested spill preparedness and response capabilities. Government and industry plans, individually and collectively, proved to be wholly insufficient to control an oil spill of the magnitude of the Exxon Valdez incident. Initial industry efforts to get equipment on scene were unreasonably slow, and once deployed the equipment could not cope with the spill. Moreover, the various contingency plans did not refer to each other or establish a workable response command hierarchy. This resulted in confusion and delayed the cleanup.

Prepared by the National Response Team, this report was requested by the President and undertaken by Secretary of Transportation Samuel K. Skinner and Environmental Protection Agency Administrator William K. Reilly. The report addresses the preparedness for, the response to, and early lessons learned from the Exxon *Valdez* incident. The President has also asked Secretary Skinner to coordinate the efforts of all federal agencies involved in the cleanup and Administrator Reilly to coordinate the long-term recovery of the affected areas of the Alaskan environment. These efforts are ongoing.

While it remains too early to draw final conclusions about many spill effects, the report addresses a number of important environmental, energy, economic, and health implications of the incident.

The lack of necessary preparedness for oil spills in Prince William Sound and the inadequate response actions that resulted mandate improvements in the way the nation plans for and reacts to oil spills of national significance.

This report starts the critical process of documenting these lessons and recommending needed changes to restore public confidence and improve our ability to plan for and respond to oil spills. The following points deserve special emphasis:

1. Tvizirxnsr$mwxli$jmrwx$pmri$sj$hijirwi2 Avoidance of accidents remains the best way to assure the quality and health of our environment. WC must continue to take steps to minimize the probability of oil spills.

62$Tvitevihriw$qywfiwxirkxlirih2 Exxon was not prepared for a spill of this magnitude--nor were Alyeska, the State of Alaska, or the federal government. It is clear that the planning for and response to the Exxon *Valdez* incident was unequal to the task. Contingency planning in the future needs to incorporate realistic worst-cast scenarios and to include adequate equipment and personnel to handle major spills. Adequate training in the techniques and limitations of oil spill removal is critical to the success of contingency planning. Organizational responsibilities must be clear, and personnel must be knowledgeable about their roles. Realistic exercises that fully test the response system must be undertaken regularly. The National Response Team is conducting a study of the adequacy of oil spill contingency plans throughout the country under the leadership of the Coast Guard.

72$V iwtsrwi$etefmpmxmiwqywfi$irlergih$s$vihygi irznvsrq irxep$vmwo2 Oil spills--even small ones-- are difficult to clean up. Oil recovery rates arc low. Both public and private research are needed to improve cleanup technology. Research should focus on mechanical, chemical, and biological means of combating oil spills. Decision-making processes for determining what technology to use should be streamlined, and strategies for the protection of natural resources need to be rethought.

82$Wsq i$smp$wtmppw$q e}fimrizmefpi2 Oil is a vital resource that is inherently dangerous to use and transport. We therefore must balance environmental risks with the nation's energy requirements. The nation must recognize that there is no fail-safe prevention, preparedness, or response system. Technology and human organization can reduce the chance of accidents and mitigate their effects, but may not stop them from happening. This awareness makes it imperative that we work harder to establish environmental safeguards that reduce the risks associated with oil production and transportation. The infrequency of major oil spills in recent years contributed to the complacency that exacerbated the effect of the Exxon *Valdez* spill.

9² Pikmₓₑₓₙsr$ sr$ pₑfₙₚₓ} $ erh$ gsqₜirwₑₓₙsr$ₙw riihih2 The *Exxon Valdez* incident has highlighted many problems associated with liability and compensation when an oil spill occurs. Comprehensive U.S. oil spill liability and compensation legislation is necessary *as* soon as possible to address these concerns.

:2§Xli$Yₙₓih$Wₓₑₓiw$wlsyₚh$vₑₓₙj}$ₓli$Mₙₓivₙₑₓₙsₙₑp Qₑvₙₓₙqie$Svkₑₙₙₑₓₙsr$,MQS-$5=<8$Tvsₓsgspw2 Domestic legislation on compensation and liability is needed to implement two IMO protocols related to compensation and liability. The United States should ratify the 1984 Protocols to the 1969 Civil Liability and the 1971 Fund Conventions. Expeditious ratification is essential to ensure international agreement on responsibilities associated with oil spills around the world.

;2§Jihivₑp$ₜₑrₙₙrk$jsv$sₙp$wₜₙₚₚw$qₙyw$fi$ₙₙq tvszih2 The National Contingency Plan (NCP) has helped to minimize environmental harm and health impacts from accidents. The NCP should, however, continue to be reviewed and improved in order to ensure that it activates the most effective response structure for releases or spills, particularly of great magnitude. Moreover, to assure expeditious and well-coordinated response actions, it is critical that top officials--local, state, and federal--fully understand and be prepared to implement the contingency plans that are in place.

<2§Wₓyhmiwsjₓli$psrk1xivq $irzₙvsrqirₓep$erh lieₚ1$ijjigₓw$qₙywfiyrhivₓeoir$i|tihₙₙsywp} erh$geviↄypp}2§Broad gauge and carefully structured environmental recovery efforts, including damage assessments, are critical to assure the eventual full restoration of Prince William Sound and other affected areas.

I. INTRODUCTION AND BACKGROUND

At 0004 on March 24, 1989, the 987-foot tank vessel *Exxon Valdez* struck Bligh Reef in Prince William Sound, Alaska. What followed was the largest oil spill in U.S. history: over ten million gallons of crude oil flooded one of the nation's most sensitive ecosystems in less than five hours. The oil slick has scattered over 3,000 square miles and onto over 350 miles of shoreline in Prince William Sound alone. The initial response was slow and insufficient. Major steps have now been taken to clean up the spill, and these efforts will continue throughout the summer.

The purpose of this report by the National Response Team (NRT) is to address preparedness for, the response to, and the early lessons learned from the *Exxon Valdez* oil spill. It was requested by the President and has been undertaken at the direction of Secretary of Transportation Samuel K. Skinner and Administrator of the Environmental Protection Agency William K. Reilly. The 14 agencies comprising the NRT worked together closely in preparing this report. Except where otherwise stated, the report covers the period from the incident through April 26.

The report describes the status of preparedness and response actions taken in the month after the incident. Preliminary environmental, energy, economic, and health effects of the spill are discussed. Preliminary recommendations for follow-up steps to prevent similar spills are identified. The report represents an important first step in examining the spill and improving both preparedness and response capabilities in the future through steps such as research and enhancement of liability and compensation provisions.

Concurrently, other studies of the *Exxon Vazdez* incident are being undertaken, and other reports will follow. The National Transportation Safety Board, Coast Guard, State of Alaska, and other authorities are looking into the spill. This report complements these efforts. Together, they will help to provide a complete picture of the oil spill's causes, its effects, and needed follow-up actions.

In his statement of March 30, the President described the *Exxon Valdez* oil spill as 'an environmental tragedy." The incident has both short-term and long-term implications. Prince William Sound is a region rich in biological diversity, and the oil spill has caused ecological harm. The spill has affected directly the livelihoods of many Alaskans. It also has impaired the beauty of a spectacular wild area that has provided inspiration not only to those persons who live and work along its shores, but also to the growing numbers of people from the rest of Alaska and elsewhere who enjoy its recreational opportunities. Alaska represents a last unspoiled frontier in the eyes of many Americans. That an incident like this oil spill can cause such damage in such a short time is a frightening realization.

Another reality is the fact that both Alaska and the rest of the United States depend on Alaskan oil. Although the reduction in oil pipeline throughput resulting from the spill was of relatively short duration, the interruption of Alaskan crude oil created serious concern regarding future supply curtailments. Americans consume about 700 million gallons of oil daily. Alaskan oil helps to limit the country's balance of trade deficit, and its steady supply also plays a role in protecting national security. In addition, oil revenues account for over 80 percent of Alaska's state income. The oil industry provides many Alaskans with jobs.

The Exxon *Valdez* incident therefore dramatizes the difficult decisions that must be made in balancing environmental protection and economic growth in Alaska. In the words of former Governor of Alaska Jay S. Hammond, 'We are called upon at once to be oil barrel for America and national park for the world.' On one hand, the Alaska National Interest Lands Conservation Act of 1980 demonstrates the priority given to preserving Alaska's natural beauty by setting aside over 100 million acres as national conservation units. On the other hand, the U.S. Government has taken steps to develop Alaska's vast energy reserves. The Trans-Alaska Pipeline Authorization Act provided the means by which oil could be transported to the Port of Valdez for shipment to the lower 48 states. The truth brought to bear by the *Exxon Valdez* incident is that accidents can occur that threaten the coexistence of conservation and energy interests.

Actions have been taken to decrease the probability of such accidents, to prepare for them, and to mitigate their impacts if they occur. More, however, can and must be done. The many possible causes of an accident make prevention difficult. These causes include: mechanical or structural failure, human error, acts of God, inadequate or inappropriate design, and sabotage.

Once the *Exxon Valdez* spill occurred, a number of circumstances combined to complicate the response action. That the spill took place in a remote location complicated an expeditious and effective response. The sheer size of the spill, which was larger than contingency planning had anticipated, posed particular problems. The magnitude of the spill was beyond the physical capability of skimmers and booms currently being used in the United States. Moreover, the first equipment to control the spill arrived on scene over ten hours after the incident after more than 10 million gallons of oil already were in the water.

A number of contingency plans were in place. Alyeska had a contingency plan. National, regional, and local plans mandated by federal regulation all had been developed. These contingency plans served as the basis for response actions.

In the absence of realistic worst-case scenarios and without adequate booms and barges on hand to contain the spill, however, these plans had an unreal quality and arguably served to reinforce a dangerous complacency.

It also is important to remember that the isolated location minimized human health impacts. With respect to environmental impacts, nature's remarkable resilience has enabled eventual recovery from large oil spills in other areas in the past.

The next step is to evaluate the adequacy of these contingency plans, both with respect to their specific requirements and their implementation. In the aftermath of the *Exxon Valdez* incident, questions have been raised about contingency planning requirements for oil spills in general and about appropriate spill liability and compensation provisions. More study of long-term environmental effects associated with persistence of oil in Prince William Sound and its implications for the food chain are needed. Research and development to improve response capabilities should be fostered. This report begins the task of investigating these and other issues raised by the *Exxon Vuldez* oil spill.

Alaska has been called upon to be both a source of energy for America and a seemingly endless frontier where nature is preserved. The difficult job ahead is to maintain this delicate balance.

II. THE INCIDENT

The American-registered motor tankship *Exxon Valdez* departed the Alyeska marine terminal in Valdez, Alaska, on the evening of March 23, 1989. The vessel was under the control of its captain, the guidance of an Alaska state pilot, and monitored by the U.S. Coast Guard Vessel Traffic Service (VTS). The VTS monitors moving vessels by radar from Valdez and Potato Point. The ship was enroute to Los Angeles/Long Beach and was loaded with 53,094,510 gallons of Prudhoe Bay (North Slope) crude oil.

The *Exxon Valdez* is a two-year-old tankship of single skin, high-strength steel construction. It is 987 feet long, 166 feet wide and 88 feet deep. The ship weighs 213,755 deadweight tons and has 11 cargo tanks. In lieu of double bottoms, the ship has seven protectively-located segregated ballast tanks (see Figure 2).

At 2325, the captain advised the VTS that the pilot had departed. He further stated that the ship probably would leave the outbound traffic lane and cross the separation zone into the inbound lane in order to avoid ice. The next call from the ship stated that it was reducing speed to 12 knots to wind its way through some ice and that the VTS would be advised after the ice had been cleared.

The *Exxon Valdez* ran aground on Bligh Reef, Prince William Sound, Alaska, four minutes after midnight on Good Friday morning, March 24, 1989 (see Figure 3). At the time of the grounding, the *Exxon Valdez* was loaded to a draft of 56 feet. The charted depth where the vessel grounded was 30 feet at low tide. The severity of the grounding is attributed to the sound's rocky bottom, coupled with the vessel's momentum. Subsequent damage surveys showed that eight of the 11 cargo tanks, extending the full length of the vessel, were torn open. Three salt-water ballast tanks also were pierced. A total of 11 tanks on the center and starboard side of the vessel were damaged.

The enormous damage caused a rapid loss of cargo. Within five hours, 10.1 million gallons had been spilled. About 80 percent of the ship's cargo remained on board, however, and the vessel came to rest in a very unstable position. The Exxon *Valdez* was in danger of capsizing if it floated off the reef. Both oil spill response and removal of the remaining oil from the ship became top priorities.

FIGURE 1
Exxon Valdez

Source: EPA, 1989.

This report is not intended to replace any investigations concerning the *Exxon Valdez* oil spill currently being conducted under existing federal or state statutes. Certain aspects of the incident therefore are the responsibility of the following parties:

o The National Transportation Safety Board (NTSB) is conducting a fact-finding investigation that can be expected to identify responsibility for the *Exxon Valdez* oil spill, as well as to make recommendations for avoiding such accidents in the future. An NTSB report typically requires eight to 12 months to prepare.

o The U.S. Coast Guard is investigating the role of the VTS and will address any circumstances that may have played a part in the spill.

o The U.S. Coast Guard also is conducting a marine casualty investigation that will examine any evidence of wrongdoing on the part of the captain, officers, or crew of the *Exxon Valdez*. This investigation will determine if the Coast Guard will initiate suspension and revocation proceedings against the licenses or documents of personnel aboard the ship.

o The Federal Bureau of Investigation is investigating whether there were criminal violations of the Clean Water Act and other applicable federal statutes.

o The Attorney General for the State of Alaska is investigating issues of negligence and liability resulting from violations of state laws.

Appendix A contains a chronology of events.

FIGURE 2
Schematic of *Exxon Valdez* Showing Damaged Tanks

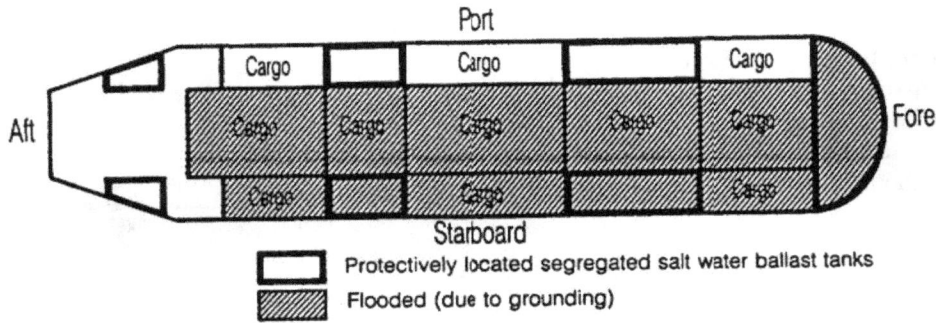

Port

| | Cargo | | Cargo | | Cargo |

Aft

Cargo | Cargo | Cargo | Cargo | Cargo

Fore

Cargo | Cargo | Cargo

Starboard

☐ Protectively located segregated salt water ballast tanks

▨ Flooded (due to grounding)

FIGURE 3
Prince William Sound

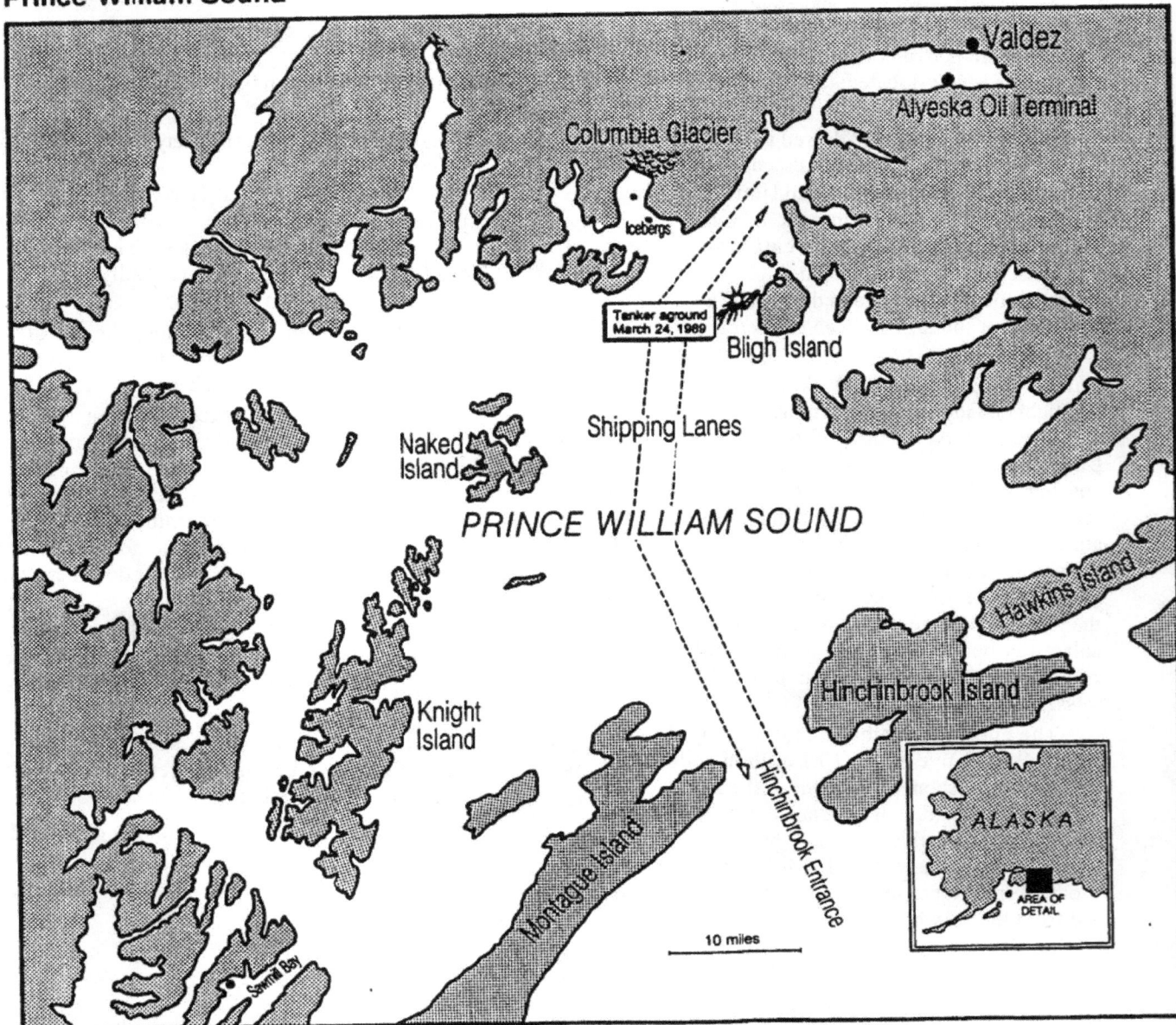

Valdez

Alyeska Oil Terminal

Columbia Glacier

Icebergs

Tanker aground
March 24, 1989

Bligh Island

Shipping Lanes

Naked
Island

PRINCE WILLIAM SOUND

Hawkins Island

Knight
Island

Hinchinbrook Island

Montague Island

Hinchinbrook Entrance

Sawmill Bay

10 miles

ALASKA

AREA OF
DETAIL

A. PREPAREDNESS

When the tanker *Exxon Valdez* ran aground on Bligh Reef, six contingency plans were in place, ranging from the National Contingency Plan (NCP) to site-specific plans for Prince William Sound. They were designed to bring about an effective and coordinated national, regional, state, local, and industry oil spill response effort. The NCP and the Alaska Regional Oil and Hazardous Substances Pollution Contingency Plan (RCP) established federal responsibilities for response and identified the Coast Guard as On-Scene Coordinator (OSC). The Alaska State Oil and Hazardous Substances Pollution Contingency Plan outlined the state role.

Initial responses were identified in both the local Coast Guard Marine Safety Office (MSO) plan for the Port of Valdez and industry's Alyeska Contingency Plan for Prince William Sound. The Alyeska plan guided the *Exxon Valdez* spill response before the Exxon Company took responsibility on March 25 for the incident and put its own plans into effect.

Oil spill preparedness is a constantly evolving process of incorporating lessons learned from simulated spills and actual incidents. Contingency planning grows from this continuing distillation of experience, shaping new requirements for response training, drills and exercises, equipment, and other resources.

1. National Response System

Composed of 14 federal agencies, the National Response Team (NRT) has broad responsibilities for the coordination of federal planning and preparedness. The NRT provides national support for response actions related to oil discharges and hazardous substance releases. Primarily through Regional Response Teams (RRT), the NRT supports emergency responders at all levels by means of technical expertise, equipment, and other resources. It also assists in the development of training, coordinates responses with neighboring countries, and manages the National Response System (NRS).

The umbrella NRS includes key elements of any federal response effort. Legally, the NRS functions under the NCP which, in turn, was created by the Clean Water Act (CWA). The Comprehensive Environmental Response, Compensation, and Liability Act (CERCLA/Superfund) also is implemented through the NCP. Section 311(k) of the CWA provides a fund for federal responses to oil spills.

Persons involved in a spill or release of more than a 'reportable quantity" of oil or hazardous substance (as defined by Superfund) are required by law to notify the National Response Center (NRC) in Washington, D.C. immediately. Staffed by Coast Guard personnel and funded by the Department of Transportation and the Environmental Protection Agency (EPA), the NRC provides a central location for reporting spills of oil and hazardous substances.

When not convened for a specific incident, the RRT is a standing body responsible for maintaining up-to-date regional planning and preparedness. The 13 standing RRTs (one for each of the 10 federal regions, plus one each for the Pacific Basin, Caribbean, and Alaska) under the NRT are essential to effective federal-state coordination in any oil spill response. Working closely with state governments and federal OSCs in its region, the RRT ensures that appropriate federal agencies provide assistance at spill scenes when federal help is requested.

Typically, the 'incident-specific' RRT consists of selected federal agency and state representatives who have technical expertise or contacts needed by the OSC for a particular incident. Depending on OSC needs, incident-specific RRT members may provide technical advice or actual resources such as equipment or manpower needed on scene. The RRT also serves as an information conduit for federal agency field offices and state staffs. Each RRT develops a Regional Contingency Plan (RCP) to delineate clearly roles and responsibilities at all levels of government during a response.

An RRT will review OSC spill reports to identify problems in regional response capabilities and help OSCs develop contingency plans for specific areas in its region. Also, RRT member agencies may provide training for contingency planners and conduct simulation exercises of regional and OSC contingency plans to test federal response capabilities.

A detailed report on the National Response System is provided in the NRT's 1988 annual report.

2. Contingency Plans

The following section describes contingency plans identifying the responsibilities of response personnel in the event of an oil spill in Prince William Sound.

a. The Alyeska Plan

The Alyeska Pipeline Service Company's oil spill contingency plan for the pipeline, terminal and Prince William Sound is an industry plan required under state law.   The plan incorporates federal requirements applicable to the terminal and pipeline. For tankers, the plan includes "General Provisions" covering Alyeska's oil spill response capability for terminal and tanker operations, a more detailed "Port of Valdez" section, and a 'Prince William Sound' section which addresses tanker operations from the terminal to Hinchenbrook Entrance. The Alyeska plan incorporates a Prince William Sound plan developed specifically for rapid and effective responses to spills from vessels in trade with Alyeska's Valdez Terminal.

The plan states that Alyeska will "direct cleanup operations of spills" from tankers carrying Trans-Alaska Pipeline System (TAPS) oil through Prince William Sound in such a way as to make federal or state intervention or takeovers unnecessary. It describes equipment and lays out procedures for oil spill detection and assessment, emergency notification and coordination, and control actions covering cleanup, disposal, and restoration.

The Alyeska plan gives priority to containment and cleanup of oil spills to prevent or minimize the amount of oil reaching sensitive areas. The plan lists 136 sensitive areas to be protected in and around Prince William Sound. The Alyeska plan addresses responseactions, reconnaissance, exclusion booming sites, response times (including a five-hour objective for initial spill response), oil transfer activities, spill trajectories, climate, oceanography, and fish and wildlife resources.

The plan covers scenarios for three spill sizes, including an 8.4 million gallon spill in Prince William Sound remarkably similar to what actually occurred on March 24. This scenario estimates that approximately 50 percent of the oil would be recovered at sea either directly after the spill or at a later time.   The Alyeska plan also establishes public relations guidance and sets forth a training program and an annual full-scale, company-wide field exercise.

Chain of command responsibilities in the Alyeska plan generally parallel the NCP. The Alyeska Oil Spill Coordinator (AOSC) heads the Oil Spill Task Force responsible for providing response and follow-up activities for all oil spills. The terminal superintendent heads reconnaissance and directs response supervisors. The immediate response teams 'evaluate a spill and begin cleanup. If, at that stage, the spill cannot be contained, additional resources are requested from the AOSC, who can activate any or all remaining task force resources.

In addition to other response strategies, the Alyeska plan addresses use of dispersants, conforming with the NCP restrictions. The plan acknowledges the need to receive approval from state and federal governments before dispersants can be applied to a spill.   However, the Alyeska plan concludes that, "it would be difficult to effectively apply dispersants..." because of the need to bring equipment in from Arizona and because of the approval process. Nevertheless, the plan emphasizes the use of dispersants as an option for spill management.

b. Captain of the Port (COTP) Prince William Sound Pollution Action Plan (OSC Plan)

This plan implements provisions of the NCP and the Alaska Regional Oil and Hazardous Substances Pollution Contingency Plan (RCP). It takes into account the Alyeska plan for Prince William Sound and the Port of Valdez.

The plan provides information on port area geography, including a general description of Prince William Sound and a more detailed discussion of the Port of Valdez and Orca inlet. It provides a listing and brief description of waterfront facilities in the Captain of the Port (COTP) zone. Procedures for notifying local, state, and federal agencies, including the Alaska Department of Environmental Conservation and other RRT members, are addressed. The plan discusses planned response actions for oil spills in five areas of the port: Hinchinbrook Entrance, Central Prince William Sound, Valdez Arm and Narrows to Middle Rock, Middle Rock to Port Valdez, and the remainder of the sound. A COTP Valdez response organization with a description of duties for each billet is provided, and procedures for gaining access to response resources (including special forces) are addressed. A list of federal, state, and local response resources extends from the NRT and Strike Teams to the Homer Harbor Master. The plan cites access to a computer listing of clean-up equipment in Alaska.

c. National And Regional Contingency Plans

Both the Alyeska and COTP Prince William Sound plans spell out a multitude of site-specific response recommendations and directives. They operate in the context of national and regional response policies established through the NCP and RCP. The principal task of the NRT is to coordinate all federal oil spill response actions and policies. Through federal regulation, the NCP provides guidance for the more site-specific RCP and OSC contingency plans.

The RCP is the region-specific plan that establishes a regional response team. It predesignates federal OSCs and outlines all regional mechanisms for coordinated response activities involving federal and state personnel. The OSC monitors the spiller's (the responsible party's) activities to terminate, contain, and remove an oil discharge. The OSC manages a response action when the spiller is unknown or unable to provide a response.

The RCP, revised in 1988, establishes a RRT for the area that includes Prince William Sound. The plan outlines the regional response system that establishes mechanical oil removal as a primary spill response strategy. In one of the first planning efforts of its kind, the RCP includes chemical dispersant preauthorization procedures for use in Prince William Sound. Preauthorization by the RRT, which is provided for in the NCP, is meant to assist the OSC in making timely dispersant-use decisions by providing him with authority to allow dispersant use in specified areas without RRT concurrence. Where preauthorization is not granted, the OSC must first obtain the concurrence of the RRT. The RCP. also contains wildlife protection guidelines for an oil spill.

Like most regional plans, the RCP provides a mechanism for coordination of state and federal assistance after a spill in support of the federal OSC's basic responsibility to either monitor the cleanup or conduct a response. In either case, the OSC is provided technical assistance by the special forces and teams listed in the RCP. The RRT also furnishes equipment and other resources. When not convened for a specific incident, the RRT is a standing body responsible for maintaining up-to-date regional planning and preparedness.

The RRT's Fiscal Year (FY) 1989 work and training plans establish a priority listing of tasks designed to improve Alaskan regional preparedness. In FY 1989, the RRT had planned to prepare guidance to help local planners effectively use federal oil spill response resources. The RRT completed planning guidelines for the rehabilitation of oiled wildlife.

d. State of Alaska Contingency Plan And Response Program

The Alaska oil spill response program, in place since 1977 to address rapid gas and oil development in the state, is administered by the Alaska Department of Environmental Conservation (ADEC). The state response program serves a number of functions: review and approval of all oil spill contingency plans; maintenance of the state's Oil and Hazardous Substances Pollution Contingency Plan; inspection of state oil facilities; prevention and cleanup of underground spills; response and cleanup or oversight of responsible party cleanup of oil spills; and enforcement of many other authorities. The ADEC approves oil spill contingency plans for over 450 tankers, barges, onshore terminals, and offshore facilities. The ADEC does not maintain. full-time oil spill response teams or large clean-up equipment inventories.

The plan lists the U.S. Coast Guard as having 'basic investigative and enforcement responsibilities for oil spills that occur on coastal waters bordering Alaska.' It adds that the U.S. Coast Guard Captain of the Port of Valdez "is the federal OSC for the coastal waters of Prince William Sound from Cape Puget to Castle Island near Cordova." The plan establishes ADEC as the lead state agency responsible for oil spill emergencies within Alaska and its coastal waters.

Under the state plan, responses to moderate (1,000 to 100,000 gallons in coastal waters) or large (over 100,000 gallons) oil spills fall under the province of the U.S. Coast Guard or Environmental Protection Agency OSC. In these cases, the state OSC will act as an advisor to the federal OSC regarding such state issues as availability of state and local resources, assignment of priority areas for cleanup and protection, response equipment and manpower staging areas, potential disposal areas, threats to humans and wildlife habitats, adequacy of cleanup, activation of the RRT, and activation of a state-funded response.

In response to a large spill, the state OSC functions as the state representative to the RRT. In this situation, response jurisdiction is assumed by the federal government once the federal OSC arrives on scene. The state then provides the federal OSC with appropriate assistance. Specifically, Annex XVII to the state plan gives ADEC authority to respond to catastrophic oil discharges which constitute a disaster emergency under Alaska Statute 26.23.010-230. Under Alaska Statute 46/04.010, ADEC also is authorized to seek reimbursement of clean-up or containment expenses. Spillers are responsible under state law to contain and clean up any oil discharge except in cases where containment and cleanup is not technically feasible or will cause more damage than the spill itself.

Other state agencies also share oil spill response responsibilities. For example, the Alaska Department of Fish and Wildlife provides for protection of fish, game, and aquatic plant resources. Other state and local agencies regulate operations that affect health and safety.

e. The Exxon Plan

The Exxon Shipping Company Headquarters Casualty Response Plan, a voluntary document not required by federal law or regulation, establishes a company Casualty Management Team and a Headquarters Oil Spill Assistance Team. The plan made available by Exxon for purposes of developing this report defines the organization and responsibilities of each team during a marine incident, but it is not specific to any location. The Exxon plan has no specific details. It includes no explanation of any interaction with the NCP, RCP, Alyeska, state or Coast Guard plans. The Exxon plan contains no information specific to the Prince William Sound or Valdez Terminal and no equipment list other than mention of a van and sampling equipment. The plan also does not prescribe uses of booms, skimmers, and dispersants. The Exxon Plan required no approval by federal or state government.

3. Contingency Plan Findings

Contingency planning coordination:

o Government and industry plans, except the Alyeska plan, did not assume a spill of the magnitude of the *Exxon Valdez* spill and the Alyeska Plan did not provide sufficient detail to guide the response. The Alyeska plan, approved by the state, was the primary plan for purposes of direct spill cleanup involving oil from the Trans-Alaska Pipeline in the Valdez Terminal/Prince William Sound area. The Exxon

plan states that the Exxon Shipping Company is responsible for containment, cleanup, and claims settlements related to spills in the waters of the U.S. from Exxon vessels. These plans do not refer to each other or establish a response command hierarchy that would take precedence in the event of a spill either at the Valdez Terminal or in Prince William Sound.

Response planning for response:

o For the Port of Valdez, all of these plans assumed that the spiller will be the responder initially. The plans assumed that the supporting OSC, state, and RRT would evaluate response actions, providing approvals when necessary unless the spiller cannot be found. There appears to have been insufficient planning, however, to assure that either the responsible party would be able to respond effectively or, if necessary, government parties could respond to a spill of this magnitude.

Exercises:

o Although several exercises required by the Alyeska plan already have been conducted, Alyeska did not utilize critiques of these exercises adequately. A critique advanced by the State of Alaska had recommended revisions to the plan. One critique had pointed out that the Alyeska on-scene coordinator is needed at the spill site to direct and supervise clean-up operations, and to interact with on-scene government agency personnel. The Alyeska on-scene coordinator did not go with the Coast Guard to the site.

Training:

o Because planners could not anticipate the manpower needed to respond to a very large, very widespread spill, there was a lack of personnel skilled in oil spill response techniques. Valuable time was used to train inexperienced workers. In addition, some response personnel and government representatives did not fully understand the NRT/RRT structure and how it works, reducing the effectiveness of available on-scene organizations and resources through unrelated or overlapping efforts and management chains.

Supply of national significance:

o The National Oil and Hazardous Substances Pollution Contingency Plan (40 CFR 300) is adequate for handling almost all oil spills. It should be reviewed and amended, as needed, however, to ensure that it activates the most effective response structure for releases or spills of national significance.

GS XT$,S WG -$tper>

o The COTP Prince William Sound Plan considers the Alyeska plan but not the Exxon plan. Lack of coordination between the Alyeska and Exxon plans appears to have caused confusion in structuring the response to the *Exxon Valdez* incident. For example, the State of Alaska was not notified by Alyeska as required when Exxon assumed responsibility for the spill response. In addition, there was no provision for review by the federal OSC, who establishes priorities for response actions. Therefore, coordination was further limited. As it turned out, the Alyeska plan was used for immediate response and the COTP Prince William Sound plan served as a basis for guiding the actions of the OSC.

Ep}iwoe$gsrxmrkirg}$tper>

o Alyeska did not carry out the objectives of its plan to direct the spill response in a manner that ensured a rapid response and the availability of adequate and usable equipment. Nor did it provide the State of Alaska with timely information on when Alyeska had turned over its responsibility to Exxon. Alyeska was not prepared to respond to this spill.

$Iuymtq irx>

o Equipment adequate to contain and clean up the spilled oil was not available during- the initial days of the incident. This is because of the magnitude of the spill, the fact that the oil barge was not certified to receive oil and was damaged, and because equipment that would have been useful was not in the inventory, A large self-contained oil skimmer would have been useful during the spill response, notwithstanding the magnitude of the event.

V iwtsrwi$wvexikmiw>

o All the basic contingency plans--the Alyeska, Exxon, Prince William Sound, Alaska RRT and state-appear to agree that the principal response strategy is physical containment and removal, along with diversion booming to protect sensitive ecosystems. They agree that other such response strategies as chemical dispersion and in-situ burning will be employed to supplement this strategy.

o Except for the Alaska RRT plan, criteria for requesting employment of additional technologies are absent from these plans. The Alaska RRT plan contains specific criteria and a detailed checklist containing evaluation factors for determining whether to approve the use of dispersants. No similar criteria, however, have been developed for in-situ burning. Early use of dispersants in most major oil spills in coastal waters is a controversial OSC and RRT decision.

o Dispersant-use decisions, can be critical to responses. Detailed and thoughtful dispersant-use preplanning can greatly improve the technical quality of such decisions.

V iwtsrwi$mq tpiq irxexmsr>

o Although the basic response strategy is outlined clearly in the Alyeska plan, guidance to help responders implement this strategy was inadequate in the *Exxon Valdez* spill. For example, the plan identifies sensitive habitats in detail and ranks them in order of response priority. It even calculates the amount of diversion booming needed to protect these habitats. The Alyeska plan does not provide clear guidelines, however, on the manpower or equipment needed to deploy this booming or the time it would take. As a result, the amount of equipment actually stockpiled at the Valdez Terminal and elsewhere in the state was not known before the response was undertaken. Several of the plans also mention the difficulty of planning for, and responding to, spills in remote areas, but they do not identify specific measures to address this problem.

o It appears that the Alaska RRT and State of Alaska plans did not adequately consider equipment, manpower, and the logistical problems associated with such a large spill.

4. Early Lessons Learned/
Recommendations: Preparedness

Gsrxcrkirg}$tpermrk>

o Planning for a large self-contained oil skimmer and other necessary response equipment was inadequate.

o The National Oil and Hazardous Substances Pollution Contingency Plan (40 CFR 300) should be reviewed and amended as needed, to ensure that it activates the most effective response structure for releases or spills of national significance. Such releases or spills should be defined, threshold criteria should be formulated to establish when this condition exists, and a pre-established, integrated command and control mechanism should be identified to address them. This mechanism is necessary in order to utilize effectively the resources of the parties responsible for the spill, the 14 federal agencies in the NRT/RRT structure, and the affected state or states and local governments.

o The federal government should consider the extent to which contingency planning should be implemented under the CWA and other appropriate authorities in major port areas. Title 33 of the Federal Code, that addresses oil pollution prevention regulations for marine oil transfer facilities, provides a starting point for this effort. At this time, there is no specific requirement for the operators of major oil terminal facilities to develop oil spill contingency plans. The EPA and the Coast Guard do require spill prevention plans. That Alyeska is required by the State of Alaska to have such a plan is atypical. Many other states leave contingency planning to industry on a voluntary basis.

o The *Exxon Vuldez* spill and analyses of the contingency plans for Prince William Sound raise concerns about the adequacy of contingency planning in other major port areas. The President has ordered a nationwide review of contingency planning in major ports to be completed within six months. The review will be conducted by the National Response Team. The Vice Chair of the NRT, the Coast Guard representative to the NRT, will direct the review. With support from the other 13 federal agencies represented on the NRT, the Coast Guard will be the lead agency conducting the study. Regional Response Teams and Coast Guard COTP will assist in review of port plans. Key issues will include ensuring that plans in other port areas address maximum probable spill assump-

tions and associated levels of manpower and equipment. Another issue is the reinstatement of the spills inventory system, which identified the location of oil spill response equipment world-wide.

Xvemmrk>

o Training programs in oil spill clean-up techniques and responsibilities of agencies under the NCP should be reviewed and expanded to include all persons who would be relied on for response to a major oil spill. Adequate training, both in the techniques and limitations of oil spill removal and in the roles and responsibilities of the various response and support organizations under the NRT/RRT structure, is critical to the success of industry and government contingency plans. In a spill of national or region-wide significance, persons and organizations not normally involved in oil spill response may be required to function in key roles.

o There should be adequately trained personnel on hand in the event of a spill.

Exercises:

o Contingency plans are important. It is equally important to test the plans in a realistic manner to ensure an effective response. Computer-aided table-top exercises and field exercises that put stress on the response system should be expanded by the National Oceanic and Atmospheric Administration, EPA, and the Coast Guard. RRTs should ensure as a part of their planning process that contingency plans are in place for areas in which major oil spills reasonably can be expected. In addition, unannounced drills should be conducted to ensure that plans and organizations can deliver responses as planned. RRTs need to ensure that training is current, that realistic evaluations of drills are conducted, and that formal revisions of plans are required based on the results of exercises.

B. PREVENTION

Prevention is the principal defense against oil spills. The old adage that an ounce of prevention is worth a pound of cure remains valid: the best way to protect the environment is to prevent spills from occurring in the first place. This truth is vividly evident in the enormous costs of the *Exxon Valdez* oil spill.

Although both government and industry have critical roles to play, prevention of oil spills cannot be accomplished without industry action to safeguard oil transport. Industry immediately should review all elements of the oil transportation and distribution system for vulnerabilities that could lead to serious oil spills. It is industry's responsibility to ensure that facilities and equipment are properly designed, safely operated, and correctly maintained.

Industry already has taken steps in an attempt to reduce the risks of oil spills. The American Petroleum Institute has undertaken a review of industry operations and will report a recommended program in three months. The review will examine manning of ships, as well as preparedness and response issues. The major owner companies of the Alyeska Pipeline also have announced plans to improve the industry's ability to prevent oil spills in Alaskan waters.

Human error is a major factor in many types of accidents. Any comprehensive prevention program must address human error through better training and equipping of personnel, safety programs, and steps to ensure that constant vigilance is exercised by management. For example, legislation, regulation, and studies under consideration address ways to prevent drug and substance abuse in the workplace. Prevention of such abuse is especially important for vital transportation services.

Government also must take steps to help prevent oil spills. Federal and state agencies currently are conducting investigations to address the cause of the accident. These investigations will examine whether navigation controls were adequate. They will look into licensing of the captain and crew and issues of negligence, liability, and criminal violation. Results from these investigations can have an important impact on preventing future oil spills.

Prevention is a complex area, and actions to address concerns about prevention will require examination in the near future of key issues by appropriate authorities. Several areas for follow-up action appropriate by federal, state, and local agencies and by industry deserve special mention here. Following is an initial list of the prevention areas that should be considered and the agencies responsible for addressing them:

Subject	Party Responsible
Wlmt$hiwmkr$erh$gsrwxvygxmsr$(e.g., double hulls, smaller tankers)	YWGK0$mrhywxv}
Q ermmrksjwlmtw$,i2k2$licensing of officers and crew, alcohol/drug testing)	YWGK0$mrhywxv}
Wlmt$q sziq irx$,i2k2$piloting, vessel traffic control, speed limitations, navigational hazard marking)	YWGK0$mrhywxv}
Wtmpp$tvizirxmsr$tperwerhstivexmsrw$q eryepw (terminals and transfer facilities, on-board response capabilities)	ITE erhYWGK0 wxexi}osgep$ ekirgmiw0 mrhywxv}
Xvemmrksjkszivrq irx${ svoiw$mr$tvizirxmsr viwtsrwmfmpmxmiw	Jihivep3wxexi}osgep ekirgmiw
Wejix}$tvskveq w$erh$xvemmrk$jsv$stivexsvw (management oversight)	Industry

IV. THE RESPONSE

A. ASSESSMENT AND INITIAL RESPONSE

Upon notification by the *Exxon Valdez* to the Vessel Traffic Service (VTS) of the incident at 0028, the Commanding Officer, Marine Safety Office (MSO) Valdez, in his capacity as the predesignated federal On-Scene Coordinator (OSC), immediately closed the Port of Valdez to all traffic. The OSC notified the National Response Center (NRC), the state, and Alyeska of the spill. By 0100, a pilot boat with a Coast Guard investigator and an Alaska Departmentof Environmental Conservation (ADEC) representative aboard was enroute to the vessel.

At 0330, the Coast Guard investigator reported that approximately 5.8 million gallons of oil had been lost from the damaged tanks. By 0530, an estimated 10.1 million gallons of crude oil had spilled into Prince William Sound.

Alyeska accepted responsibility for the cleanup and activated its emergency operations center 45 minutes after receiving word of the spill. A second operations center in Anchorage was established 15 minutes later.. By 0500, 70 people had been called, and 28 people were working at the terminal.

By 0730, Alyeska had a helicopter aloft with a Coast Guard investigator aboard. Videotape recorded during this overflight showed a slick 1,000 feet wide by about four miles long.

B. RESPONSE

1. Strategy

Several goals were developed by the responsible party and the OSC in structuring the response. Of primary importance were steps to prevent the spill of additional oil. The *Exxon Valdez* was unstable. Genuine concern was expressed that the vessel could capsize and spill the 43 million gallons of oil remaining on board (80 percent of the total load) into the water.

Environmentally sensitive areas needed protection. The National Oceanic and Atmospheric Administration (NOAA) had identified these areas prior to the incident. This information was used to set priorities for sensitive areas, such as fish hatcheries. Four fish hatcheries were singled out as priorities, and equipment was deployed to protect them. The OSC faced the decision to use the limited available booms and skimmers to protect other sensitive areas or to contain the oil, because not enough equipment was on hand to do both tasks.

Personnel safety was another major concern. Flammable, toxic fumes made response actions dangerous. To address this danger, Exxon established a nine-member health and safety task force, developed a safety plan, and trained all personnel on scene. No one was seriously injured in this case. Finally, strategies were developed for the containment and recovery of the 10.1 million gallons of oil that had entered the environment.

The response began on several fronts. The OSC mobilized the National Strike Force to assist with controlling the release of additional oil from the *Exxon Valdez* Personnel from the MSO were recalled and the pre-arranged response organization put in place. Exxon activated its emergency center in Houston and mobilized equipment to stabilize the stranded vessel. The *Exxon Baton Rouge* was directed to the scene to offload the remaining cargo. Within 24 hours of the incident, the oil on the *Exxon Valdez* was being pumped off.

Major problems began to arise, however, within three hours of the incident. Alyeska response equipment shortages and casualties delayed any opportunity to contain the spilled oil early on. A barge specified by Alyeska in its contingency plan to provide containment equipment at the scene of an oil spill within five hours of notification had been stripped of its equipment for repairs. Reloading the barge was time consuming and further delayed when the cranes loading the equipment on the barge were redirected by Alyeska to load transfer equipment aboard a tug. The barge left for the vessel ten hours after the spill and arrived on scene two hours later.

2. Logistics

The spill occurred in a remote location of the country, and, as the oil spread, it moved to even more difficult and remote areas. The town closest to the spill site is Valdez, which has less than 4,000 residents. The small airstrip in Valdez normally handles about 10 flights a day, but this figure jumped to between 750 and 1000 flights a day during the spill response. The Federal Aviation Administration set up and manned a temporary air traffic control tower to manage this increase in traffic. The largest plane Valdez can accommodate is a C-130 with load restrictions.   Larger planes carrying clean-up equipment were forced to fly into Anchorage. The drive from Anchorage can take up to nine hours by truck, and the roads sometimes are closed due to bad weather and avalanches. The only other town close to the oil spill is Cordova, which can be reached only by boat or plane. Its only ferry has been used as part of the response action.

Valdez has limited accommodations and could not handle the number of spill response personnel, government officials, and reporters who descended on the town in the spill's wake.   Because of a limited phone system, during the first week after the spill several thousand calls into or out of Valdez were not connected.   This lack of phone lines initially delayed responders from requesting the resources they needed to handle the spill. Only the Coast Guard OSC had a direct line out of Valdez. The phone problem was alleviated somewhat during the second week when the phone company increased the number of lines into town.

Even when equipment and clean-up personnel arrived in Valdez, the spill site still was two hours distant by boat. By April 13, the oil covered 1,000 square miles and reached 80 to 90 miles from Valdez.   It took eight to 10 hours by boat, at 10 knots, to go from one end of the spill to the other. It took 14 hours to tow a skimmer 35 miles across the width of Prince William Sound. It was an hour flight by helicopter to reach affected beaches. Staging had to be done on scene from mobile platforms, requiring that equipment be air-dropped or delivered by boat. All of these factors exacerbated the slow delivery of clean-up equipment.

Radio transmissions cannot travel great distances without repeaters in mountainous terrain. The large number of boats involved in clean-up activity (see Figure 4) resulted in multiple, simultaneous transmissions and led to many radio channel violations. This problem was eliminated when the Coast Guard cutter *Rush* arrived on scene and began enforcing proper radio procedures.

FIGURE 4
Total Number of Vessels On Scene

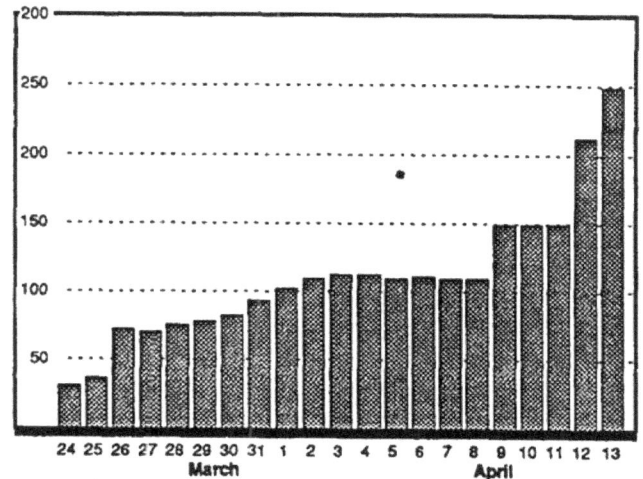

Source: U.S. Coast Guard, 1989.

3. Exxon Actions

Exxon Shipping Company assumed financial and clean-up responsibility from Alyeska on the second day after the spill; part of the normal response process and not unanticipated. Exxon designed a plan to transfer the remaining 43 million gallons of oil and salvage the vessel with the assistance of the National Strike Team.   The oil was removed by April 4 without further damage to the vessel, additional spillage, or injuries.   The vessel was' refloated and towed to a sheltered harbor for damage assessment and temporary repairs the next day.

Exxon also initiated a communications network. Exxon built a remote transmitter by day four of the spill, and on day six sent a communications trailer out on a boat. Eventually, a complete communications network was established. It consisted of four repeater stations: one on a mountain, one on Knight Island, and two on vessels.

The weather changes quickly and varies from one part of the sound to another. Clean-up personnel had to be placed on and removed from steep rocky shores which experience 10- to 12-foot tide changes twice a day. Bad weather grounded helicopters used to transport these personnel. Vessels needed warning to move equipment to safe harbors during rough seas. The Bureau of Land Management had portable weather stations for remote locations that could relay weather information through a satellite link. Because weather forecasting was critical to the response operation, Exxon worked with NOAA to set up five weather stations around Prince William Sound.

Fuel required for transit from Valdez to clean-up sites left helicopters with only about 25 minutes' worth of fuel to remain on scene. Exxon set up a refueling station in Seward to increase on-scene time for helicopters. Questions were raised concerning the spill's effect on water quality and on the fishing industry. Exxon developed a water sampling plan and selected 25 sites to monitor on a regular basis.

The Exxon emergency center in Houston opened shortly after receiving word of the incident. By the end of the fourth day after the spill, a total of 274 tons of booms, skimmers, dispersant, and dispersant application equipment had arrived. This equipment was flown in from around the world (see Figure 5).

Exxon also mobilized a fleet of vessels. By the second day after the spill, Exxon had contracted for 52 vessels. This steady build up continued until April 13 when the number of vessels hired leveled off at about 110. As of April 13, Exxon had 248 vessels contracted (see Figure 6). Eighteen contract aircraft were on site by April 13. By April 13, aircraft had delivered 1,767 tons of equipment, and twenty-five aircraft were contracted for use on site. As of April 10, the response staff totaled 1,470 and included 130 Exxon employees, 962 contract personnel, and 378 fishermen (see Figure 7).

FIGURE 5

Air Freight Delivered by Exxon

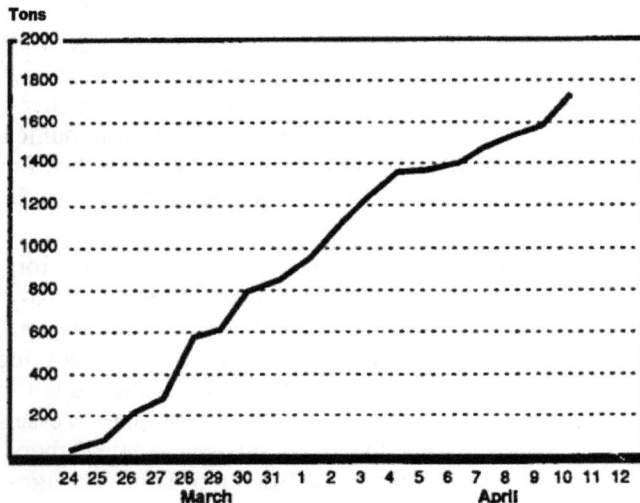

Source: U.S. Coast Guard, 1989.

FIGURE 6

Exxon Equipment On Scene

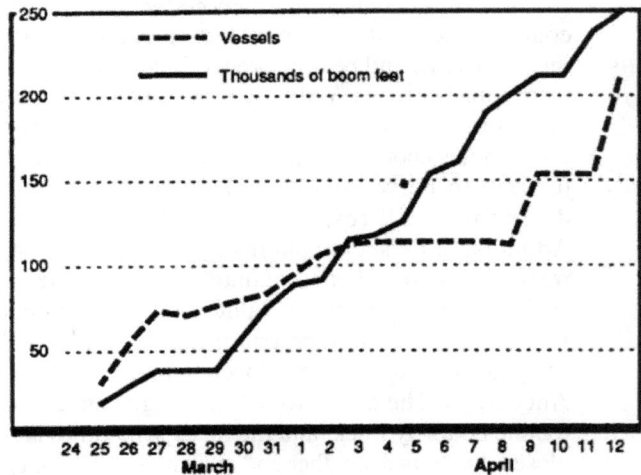

Source: U.S. Coast Guard, 1989.

FIGURE 7

Exxon Personnel On Scene

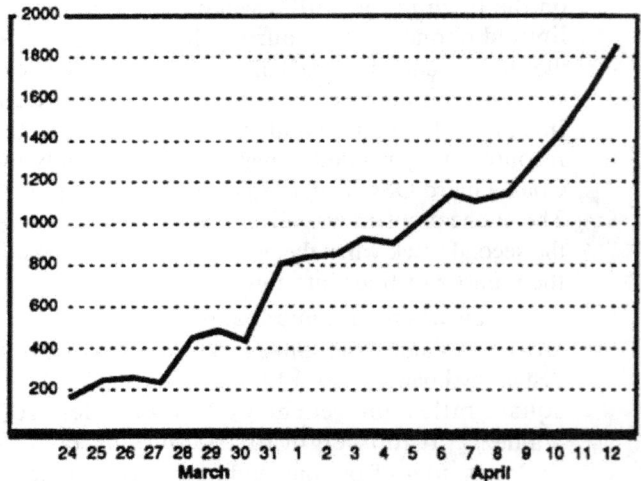

Source: U.S. Coast Guard, 1989.

4. OSC Actions And Activities In Support Of The OSC

Within 24 hours of the spill, the Coast Guard had supplemented the 34 personnel assigned to MSO Valdez. Four members of the Coast Guard's Pacific Area Strike Team (PACAREA) of the National Strike Force arrived on scene from San Francisco, and the Coast Guard began to mobilize over 1,000 personnel. The following day, the Coast Guard mobilized several cutters, aircraft, and other equipment (see Figure 8). As of April 24, over 250 personnel were working for either MSO Valdez or MSO Anchorage, and 750 were crew aboard the nine Coast Guard cutters and eight aircraft (see Figure 9).

FIGURE 8

Federal Equipment On Scene

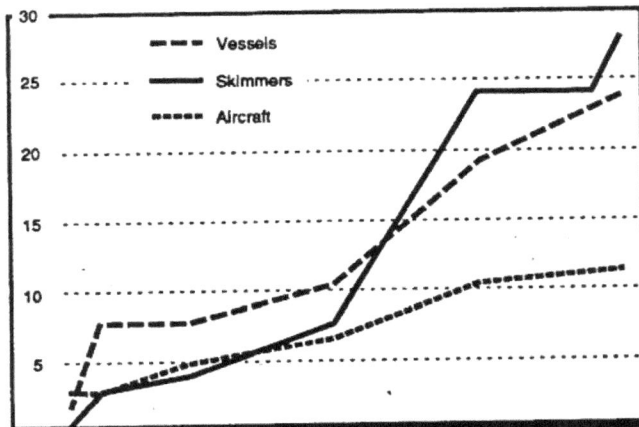

source: U.S. Coast Guard. 1989.

The National Strike Team supervised the off-loading of the Exxon Valdez under the direction of a salvage master. At any given time during the crisis, the Strike Team has had at least 20 members on scene. Virtually all of the PACAREA Strike Team members have participated in this spill response, as have some Atlantic Area (LANTAREA) members.

FIGURE 9

Federal Personnel On Scene

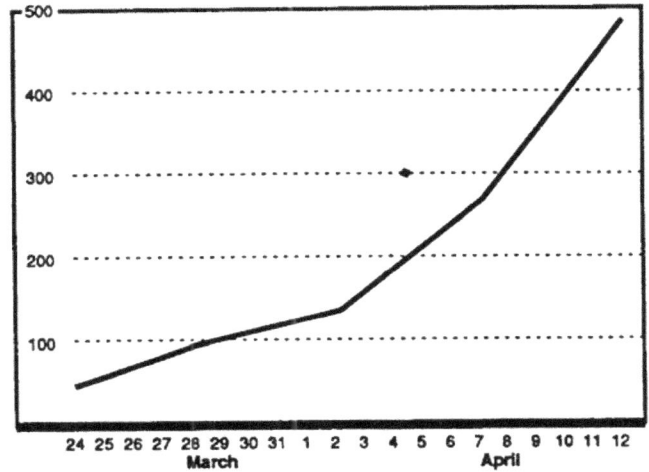

Source: U.S. Coast Guard, 1989.

NOAA's Scientific Support Coordinator (SSC) in Anchorage was called the morning of the incident and was on scene within hours. By the end of the day, NOAA's Hazardous Materials Response Branch had assembled a team of six people and a NOAA helicopter to assist the OSC Oil spill trajectory, weather, and tide forecasts were available to the OSC within the first 24 hours, and routine oil spill overflights to map the extent of contamination began on March 26. By the middle of April, NOAA had staff in Valdez, Whittier, Seward, Homer, and Kodiak to facilitate the coordination of scientific activities.

Shortly after the incident, Alyeska contacted the International Bird Rescue Research Center (IBRRC) and placed it under contract. IBRRC staff arrived on scene March 25 and began to set up a bird cleaning and rehabilitation center in Valdez. Department of the Interior (DOI) and Fish and Wildlife Service (FWS) representatives arrived in Valdez the evening of March 24. A specialist arrived in Valdez from the Hubbs Marine Research Institute in San Diego to set up a sea otter facility. Exxon hired these animal rescue specialists pursuant to the wildlife protection guidelines of the Regional Contingency Plan (RCP).

A sea otter collection program began on March 29. The sea otter cleaning and rehabilitation center in Valdez was opened on March 30. Cleaning and rehabilitation centers also were established in other cities. In addition to sea otter collection by Exxon contractors, FWS personnel are assisting in capturing oiled sea otters.

A bird collection program began on March 29, and the cleaning and rehabilitation center in Valdez opened on March 31. As the oil progressed southward, the IBRRC established centers in other cities. In addition, Exxon is reviewing an FWS draft spring migration bird protection plan. The plan focuses on hazing birds from oil-contaminated areas and establishing flight patterns to prevent aircraft from moving birds from unoiled areas to oiled areas.

Other military resources have been mobilized to clean up the spill. The Department of Defense, through the normal RRT/NRT mechanism, provided U.S. Navy equipment on the second day of the spill and increased the level of support throughout the cleanup. The Army, Navy, and Air Force have supported the response effort. (See Appendix B for Forces On Scene). Funding for this support was provided by Exxon through a previously developed mechanism using the §311(k) Fund established by the Clean Water Act (CWA). The OSC retained responsibility for cost accounting.

Coast Guard administrative, management, clerical, and support personnel from around the country have been assigned to the spill. Coast Guard public affairs officers and specialists are in the Valdez area to answer media questions, escort VIP visitors, and administer a long-term community relations program.

In addition to the resources provided by the U.S. government, state agencies, and private domestic companies hired by Exxon, the governments and private sectors of Canada, Denmark, France, Norway, and the Soviet Union made generous offers of assistance.

The RRT became actively involved in the cleanup soon after the spill. The RRT convened for its first teleconference call at 1200 on Friday, March 24, and continued to hold teleconference calls virtually every day thereafter.

The National Response Team (NRT) was activated almost immediately. Less than four hours after the spill was reported, the Coast Guard Vice Chair, acting as NRT Chair for this incident, directed the National Response Center to brief NRT members. The Coast Guard had all of its pollution reports sent directly to individual NRT agencies as the OSC released them. Shortly after the spill, the Chair and the Vice Chair visited the spill site and returned to brief the NRT.

The NRT convened five times through April 26 in special session to receive updates on the incident and discuss the activities of NRT agencies. The NRT members established a communications network for information and coordination and volunteered their resources and expertise, as needed. The Coast Guard established a 24-hour crisis action center adjacent to the NRC. Additionally, the NRT sent a group of Coast Guard, EPA, DOI, and Department of Energy representatives to Alaska for firsthand input into this report.

5. State of Alaska Actions

The State of Alaska's participation in the response to this incident began when the spill was reported. An ADEC official went to the scene with the Coast Guard investigator less than one-half hour after notification of the spill. ADEC was an integral part of the contingency planning process before the incident and an active participant in RRT and subsequent management organizations in the early stages of the response effort.   It did, however, assume a role largely independent of the federal response organization as the cleanup proceeded. For example, Alaska obtained recovery equipment and a state ferry to use as a mobile operations center on its own.

The Governor of Alaska personally surveyed the damage by 0600 the first day and remained involved with the state's efforts throughout the response. On March 26, the Governor declared a state disaster and implemented the Alaska Emergency Plan.   The following day, the Governor requested a Presidential declaration of an emergency under Title V of the Stafford Act. Specifically, he asked for the appointment of a federal coordinator and for technical and advisory assistance to the federal, state, and local governments. On April 11, the director of the Federal Emergency Management Agency advised the Governor that the National Oil and Hazardous Substances Pollution Contingency Plan (NCP) was operating. The NCP already provided the federal coordinator and technical assistance specified in the governor's request.

6. Clean-up Methods

a. Dispersants

Dispersant use became a controversial issue in this response. The federal OSC is responsible for ensuring minimal harm to the environment. The OSC relies on advice from the Scientific Support Coordinator, recommendations from the Regional Response Team (RRT), and his best professional judgment. The Alaska RRT was well prepared to address questions of dispersant use (see Appendices D and E) and had developed a new dispersant-use plan in early March 1989 (see Figure 10). The plan was incorporated into the RCP and reflected in-depth study done in the area.

At the time of the incident, two different dispersant use strategies were in effect. Each strategy applies within discrete seasonal and geographic boundaries. The area of Prince William Sound from two miles below the grounding site and extending to the north (including Valdez Harbor) is designated in the RCP as Zone 2 from March 1st until October 15th. In Zone 2, RRT concurrence with the OSC's recommendations is required prior to dispersant use authorization. (This area is designated Zone 1 from October 16th until April 30th.) The area beginning two miles south of the grounding site, which was in the path of the slick., is designated in the RCP as Zone 1 all year. In Zone, 1, the OSC is preauthorized to use (or allow the spiller to use) dispersants, with the only condition that the EPA and state (ADEC) RRT representatives be notified retroactively-but with minimal delay.

Approximately 30 minutes after the spill was reported, the OSC contacted Alyeska to suggest it consider calling aircraft to be used for dispersant application. About 0430, the OSC discussed dispersant use with Exxon and advised that dispersants were preauthorized by the RRT at the discretion of the OSC for use in Zone 1.

The OSC contacted Alyeska at 0630 and advised Alyeska to start the dispersant-use request process. About 0830, Alyeska transmitted a IO-page formal request to have both fixed-wing aircraft and helicopters spread 50,000 gallons of dispersant, beginning at 1400 that afternoon. At the time of the request, Alyeska had less than 4,000 gallons of dispersant at its terminal, no dispersant application equipment, and no aircraft. A total of 8,000 gallons of dispersant were available in Kenai, and an additional 8,800 gallons of dispersants were available in Anchorage. Alyeska had contacted a dispersant application equipment contractor in Kenai and its contract dispersant aircraft in Arizona. The RRT discussed dispersant use during its 1200 conference call. At about 1500, by which time the oil had spread south into Zone 1, the OSC granted permission for a trial application to determine the dispersant's effectiveness under the existing conditions. Alyeska conducted the first trial application about 1800 that first evening using a helicopter and dispersant bucket. The OSC decided the application was ineffective because wave action was insufficient to mix the oil and dispersant. However, the OSC authorized additional applications in Zone 1 for the following morning using a fixed-wing aircraft.

Exxon's fixed-wing aircraft arrived in Anchorage about 0615 the following morning, March 25. At about 0945, the RRT met and discussed dispersant application in Zone 2 near the spill site. The RRT concurred with the OSC that, despite calm water and light winds which adversely affected the dispersant's effectiveness, further trial applications should be conducted. The OSC announced this authorization for trial application at about 1200. Full application in Zone 2 would depend on the state and EPA RRT representatives' concurrence with the OSC's recommendation. The second trial application was conducted by Exxon using a fixed-wing aircraft with 2,500 gallons of dispersant. The aircraft arrived at the spill site around 1700. Dwindling daylight prevented a complete evaluation, but extremely calm seas and winds of less than 15 knots caused very little dispersing action to be observed by the OSC. The OSC authorized another trial application during daylight hours for the following day, March 27.

Figure 10
Zones Of Dispersant Use In Prince William Sound

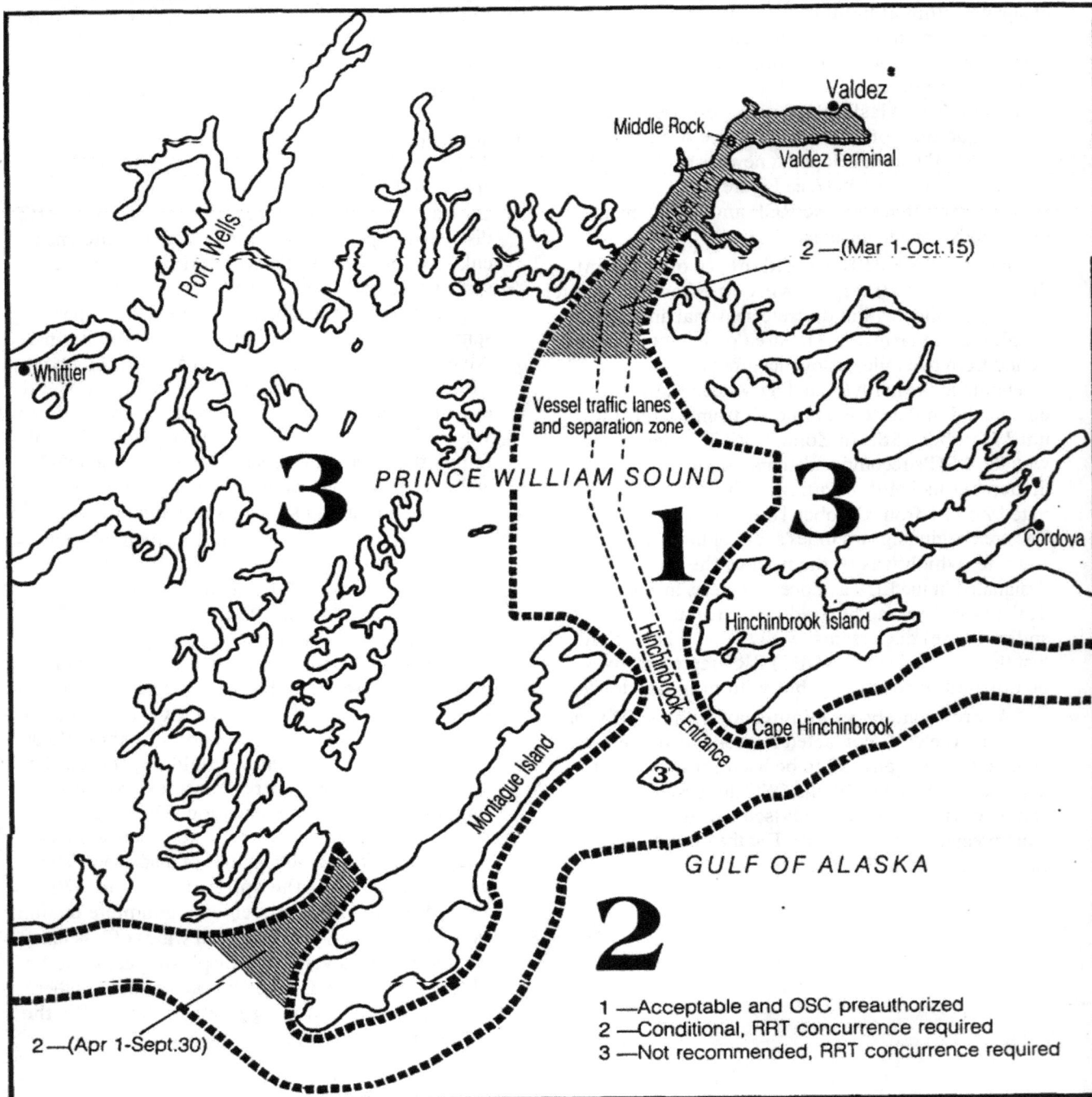

Valdez

Middle Rock

Valdez Terminal

2 —(Mar 1-Oct.15)

Port Wells

Whittier

Vessel traffic lanes and separation zone

PRINCE WILLIAM SOUND

3

1

3

Cordova

Hinchinbrook Island

Hinchinbrook Entrance

Montague Island

3

Cape Hinchinbrook

GULF OF ALASKA

2

1 —Acceptable and OSC preauthorized
2 —Conditional, RRT concurrence required
3 —Not recommended, RRT concurrence required

2—(Apr 1-Sept.30)

A third trial application was conducted about 1030 the third morning after the spill using 3,500 gallons of dispersant in a fix-wing aircraft. Mechanically, the test was not satisfactory because the nozzles did not spray the dispersant evenly. This caused some oil to be completely untreated and some oil to be overdosed with the dispersant. Throughout the day, the seas were increasing to about four feet and the wind was increasing to about 20 knots. Exxon conducted a fourth trial application at approximately 1600 that day using a different type of fixed-wing aircraft. The increased wave action created the mixing action needed to make the dispersant effective. The OSC authorized full scale use for the following day.

During the night, winds increased to gale force and continued through the following morning, grounding all aircraft. When the spotter helicopter got airborne around noon, the oil had moved into Zone 3 where dispersant use is not recommended. The RRT was contacted and asked to grant permission to apply dispersants in this "not recommended" zone. At 1400, the RRT granted permission for one plane load of dispersant to be applied in this zone. At about 1800, the OSC, along with ADEC and EPA representatives, was aboard a spotter helicopter awaiting the arrival of the dispersant aircraft. The OSC cancelled the application around 1850 when the aircraft had not arrived, daylight was waning, and there was little oil left in the area. It was later learned that, at the direction of the Exxon command post, the aircraft had applied dispersant at about 1500 over an unauthorized location.

b. Burning

On the first day of the spill, Exxon requested an open-burn permit from the State of Alaska. The state responded the following day by authorizing an effectiveness test for burning the spilled oil, and the test was conducted toward evening of that same day. Approximately 12,000 to 15,000 gallons were burned. Disagreements arose between Exxon and the State of Alaska about the success of this operation. Although the oil burned satisfactorily, there were questions about residual smoke. Some residents several miles from the burn site reported irritated eyes and throats. No further tests were conducted. The ADEC took the position that it was not opposed to burning as long as communities were not harmed and their residents were notified of an upcoming burn. The weather changed by the evening of the third day, making conditions unfavorable for another burn.

c. Mechanical Recovery

Mechanical recovery was the preferred method of oil removal because mechanical recovery removes oil from the environment without possible environmental effects from contaminants, such as added chemicals. Necessary recovery equipment included various booms, skimmers, and containment vessels. Equipment assembly was labor intensive and time consuming. Booms required personnel who could attach sections, set, and tend them. Some booms are inflatable, but one such boom sank on the first day of the spill. The booms had to be towed slowly to prevent damage. Since Prince William Sound is very large, the time necessary to relocate booms to different areas of Prince William Sound was considerable.

Skimmers are mechanical devices that remove oil from water. They require tending during operation. Skimmers must be directed to oil locations from aircraft to assure greater efficiency, thereby increasing coordination problems. Few aircraft were available initially to coordinate the deployment of skimmers. With limited personnel available to monitor and repair skimmers operating great distances from one another, long periods of inactivity resulted when they became disabled. When breakdowns required shop work, they were towed back to Valdez. For example, one skimmer with a gear box problem required 12 hours to be towed to Valdez for repairs. The repair shop was already working on two other skimmers and repairs took all night to complete.

As is graphically shown in Figures 11 and 12, 63 percent to 85 percent of the skimmers on scene were deployed at any one time after the first day. For reasons mentioned above, not all of the deployed skimmers were recovering oil effectively. A skimmer's efficiency depends on the type and condition of the oil being recovered. After the oil weathered in this spill, it had the thick, viscous consistency of axle grease. Skimmer hoses clogged, and only about 10 percent of the designed recovery rate could be achieved (see Figure 13). Heavy kelp concentrations also contributed to clogging.

FIGURE 11

Number of Skimmers On Scene

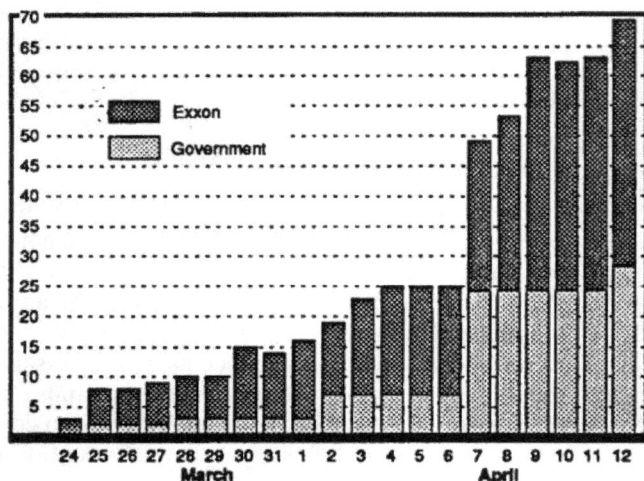

Source: U.S. Coast Guard, 1989.

FIGURE 12

Number of Skimmers in Operation

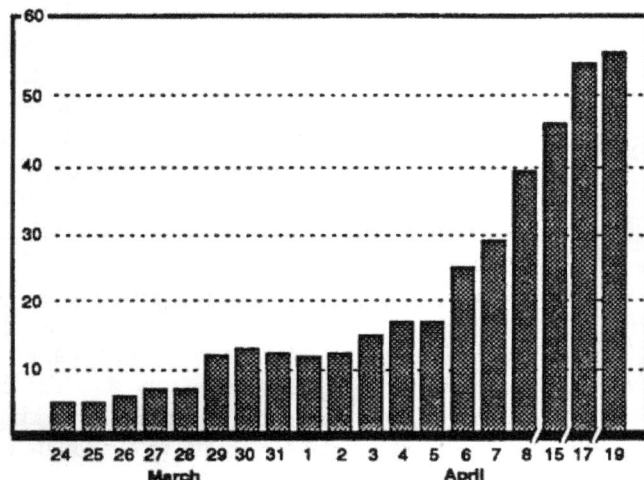

Source: U.S. Coast Guard Pollution Reports, 1989.

FIGURE 13

Estimated Cumulative and Daily Volumes of Oil Recovered

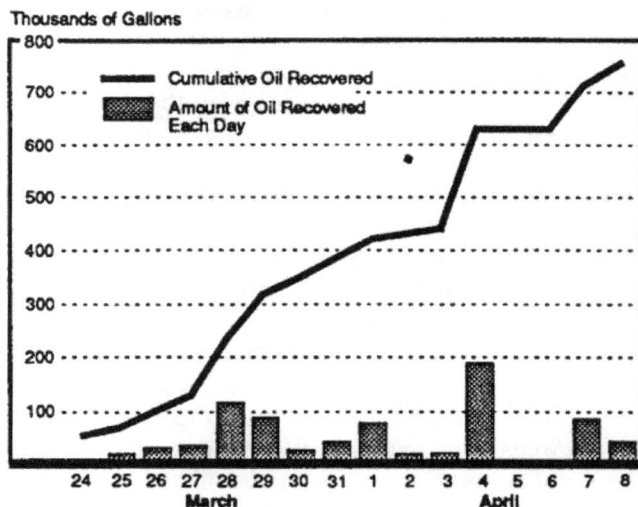

Source: U.S. Coast Guard Pollution Reports, 1989.

The third component of a mechanical recovery system is the temporary storage vessel. A small, temporary oil-containment device (oil bladder) attached to the skimmer must be emptied at a large oil recovery barge when full. This procedure was slow because the transfer pumps had difficulty moving the heavy, grease-like material. Consequently, vessels would often queue up at the recovery barge.

Weather also affected the pace and effectiveness of oil recovery. Severe weather suspended operations a number of times, forcing vessels to tow skimmers and booms to sheltered harbors and coves.

7. Shoreline Cleanup

Appropriate shoreline clean-up techniques vary according to the type of shoreline, the nature of the oil, and the natural resources present. One of the primary selection criteria in choosing a clean-up method is to ensure that the technique will not cause greater harm than allowing natural processes to cleanse the environment. Much of western Prince William Sound is characterized by high vertical cliffs frequently edged with gravel shorelines. The inner sound is laced with fjords and more sheltered rocky beaches. General recommendations for clean-up methods for the variety of shorelines found in the sound are shown in Appendix H.

After the Exxon Valdez incident, a multi-agency group was tasked to advise the OSC by identifying sensitive environments, delineating the degree of affected shoreline, and ranking areas to be cleaned. Clean-up guidelines were developed and presented to the OSC. Exxon, federal, and state agency personnel conducted field surveys to define the shoreline, determine the amount of oil present, identify logistical problems, and investigate other concerns. Cleaning techniques were identified, tested, and approved by the OSC. A streamlined process was developed by Admiral Yost in mid-April which requires OSC approval for any particular beach segment before actual cleaning work by the Exxon teams. Initially, approval of shoreline clean-up procedures was time consuming because of the large number of participants in this process with varying levels of knowledge. Special clearances remained necessary to prevent. damage to potential archaeological sites commonly found along the shores. Areas where seal pupping is expected to occur around the third week in May received high priority for cleanup.

C. RESPONSE ORGANIZATION

Under the National Contingency Plan (NCP); the OSC is responsible. for ensuring a proper response by continuously assessing and monitoring all response actions, and by 'federalizing'' a spill if the response activities are unsatisfactory. In the context of the Clean Water Act (CWA) and Executive Order 11735, "federalize" is defined as the use of federal funds for cleanup under the direction of the federal OSC. If a response is not being carried out properly, the OSC will notify the party responsible for the spill of its liability for costs associated with the removal under CWA and "federalize" the response and cleanup. Lacking a finding that a responsible party is not conducting a proper clean-up, the NCP envisages a cleanup conducted and paid for by the responsible party.

The OSC is the focal point in effectively coordinating the response to an oil spill. The OSC pulls together the various threads of expertise and provides oversight direction for the use of manpower, equipment, and resources. In general, the OSC classifies the size of the discharge; investigates the source, type, and quantity of the discharge; and monitors the response action to determine if the discharger is carrying out the response properly. In this case, the predesignated OSC was the Commanding Officer of the Coast Guard MSO in Valdez. That office consists of 34 personnel and is responsible for marine inspections, casualty investigations, port safety, environmental response, and the VTS system for the MSO Valdez area.

With Alyeska accepting responsibility for the cleanup immediately after the spill occurred, the OSC established a response organization in conformance with the NCP. The lack of preparedness on the part of Alyeska to have the requisite equipment pm-staged, however, effectively delayed any meaningful response.

The spill's sheer size and complexity of the required response taxed the initial OSC organization. Additionally, public and media concerns over the spill's potential environmental and economic effects demanded the OSC's attention to a far greater degree than that previously experienced during any spill in U.S. history. It was quickly evident that this spill would require additional assistance for the OSC.

To alleviate pressure on the OSC, the Seventeenth Coast Guard District Commander was dispatched to the scene on the second day after the spill. The presence of the District Commander at the scene may have led to initial confusion as to who was in charge, because he is the OSC's supervisor. In fact, no transfer of authority took place, and the OSC retained his role under the CWA. It may be unrealistic to believe that any disaster of national proportions such as this one will remain under the full authority of a predesignated OSC for very long.

As the response activities increased in intensity, the response organization grew to accommodate the increased demands placed on it. By the fourth day of the spill, a high level management steering committee, consisting of the Seventeenth Coast Guard District Commander, the President of Exxon Shipping, and the Commissioner of ADEC, evolved to coordinate the response. This elevation of authority was appropriate given the circumstances of the spill.

'Federalization' is a potential response in any spill. By continuously monitoring the responders-first Alyeska, then Exxon - the OSC determined, however, that the responsible party was mounting as effective a response as possible. (See Appendix B for Forces On Scene.) The OSC deemed it inappropriate to "federalize" the incident as long as Exxon continued to cooperate with the federal OSC, fund the entire operation, and perform satisfactorily.

On April 7, the President directed the Secretary of Transportation to serve as the coordinator of efforts by all federal agencies involved in the clean-up. He also directed the Commandant of the Coast Guard to return to Alaska and assume personal oversight direction of the spill response efforts. A much larger federal organization, with the Coast Guard Commander, Pacific Area, designated as federal OSC, was established.

The President directed the Secretary of Defense to make U.S. military resources available to assist in the cleanup. The Secretary of the Army was designated as the executive agent for DOD's involvement in clean-up activities, and the Director of Military Support (DOMS) was designated the action agent to coordinate, manage, and task all DOD support. To ensure coordination of requests for support, a DOMS Oil Spill Joint Task Force consisting of Army, Navy, Air Force, Marine Corps, Joint Staff, and Coast Guard representatives was activated in the Army Operations Center, Pentagon. (See Appendix B for Forces on Scene.)

D. PUBLIC INFORMATION

The Coast Guard immediately activated its existing public information plan after being notified of the spill. A local petty officer acted as interim spokesman until the District Public Affairs Officer and a Public Information Assist Team member from Headquarters arrived on scene. Their first actions were to establish a Coast Guard news office and request additional public affairs staff.

The huge number of media correspondents in the area strained the Coast Guard's ability to provide information. Phone lines were jammed as correspondents held them open to stay in contact with their home offices. Media representatives soon located themselves in a Valdez community building, where subsequently they were briefed by members of the response organization. Additional Coast Guard public affairs staff arrived five days following the spill and have run an integrated and responsive operation under extremely trying conditions.

E. EARLY LESSONS LEARNED/ RECOMMENDATIONS

1. Initial Response

o Mobilizing equipment and personnel in the initial stages of an incident is difficult in all major oil spill recovery operations. In the Exxon *Valdez* spill, this problem was accentuated by the remote location and the distances involved in moving equipment. The time lag in transporting and deploying equipment forced the responders into catch-up efforts from the onset.

o Equipment staged at the Alyeska terminal was not sufficient to cope with a spill of this magnitude. The time lag in transporting additional equipment to the scene from out-of-state led to a perception of inaction.

o Given the limitation of existing plans and capabilities, the quantity of oil released in such a short period (10.1 million gallons in five hours) overwhelmed recovery and containment efforts.

o Alyeska's initial efforts to get its equipment on scene were slow because the response barge was not ready. The response barge was stripped of equipment, took ten hours to load, and took another two hours to reach the Exxon *Valdez*. Once started, oil recovery progressed very slowly.

o The quantity and size of booms were insufficient to respond to the spill adequately.

o Few skimmers were working on scene during the first 24 hours. Alyeska also lacked a tank barge into which the skimmers could discharge recovered oil.

o The issue of dispersant use remains in dispute. Conflicting documentation from Exxon and the OSC makes the decision process unclear. It is clear, however. that neither Alyeska nor Exxon had sufficient quantities of dispersant available for the magnitude of the spill. The dispersants that were available were not used immediately because the OSC determined, as a result of three trial applications, that they were not effective due to the lack of wave action required to mix dispersants with oil, and because they were applied improperly. Dispersant use on catastrophic spills needs further study.

o Burning the oil was possible and was done. Apparently, it was not continued because of a misunderstanding between Exxon and the State of Alaska over the conditions under which burning could proceed. By the time the misunderstanding was worked out, the opportunity had passed.

2. Wildlife Rescue

o Initial wildlife rehabilitation efforts were slow. In light of the magnitude of the spill, all available resources should have been brought to the scene more quickly.

o An overall plan should have been developed during the initial stages of the spill to address bird and sea otter collection, cleaning, and rehabilitation programs that would extend throughout the entire area potentially affected. This plan is being developed now and will include volunteers and effective oversight by federal agencies responsible.

o The Alaska regional Oil and Hazardous Substances Pollution Contingency Plan's Wildlife Protection Guidelines should be expanded to include information on dealing with wildlife impacts when the responsible party assumes responsibility.

3. Response Organization

o In general, most spills are managed using a 'team' concept. This approach involves the spiller, the OSC, other federal agencies, and the state. This team concept appeared to break down into adversarial relationships that may have caused a lack of communication and ineffectiveness in the cleanup.

o The OSC spill-response organization outlined in the NCP was not followed in this case. Public and media interest required a disproportionate amount of the OSC's time.   Three different organizations eventually evolved to deal with this spill: the NCP-specified organization, the steering committee, and Presidentially-directed oversight by the Commandant of the Coast Guard.   An initial perception persisted that strong oversight direction was not being exercised.

o The NCP should be reviewed to determine the most appropriate organizational structure for catastrophic spills.

o The Commandant of the Coast Guard reestablished federal oversight of the response, clarified lines of authority and management, and put response actions back on course.

o 'Federalization' of the spill was not considered necessary because the OSC determined Exxon was taking proper action to remove the spilled oil and effective action to eliminate its threat to the environment. Exxon assumed responsibility for the spill quickly and showed the willingness and capacity to acquire the equipment and personnel necessary to carry out the response.

4. Response Policies

o When tested by a massive, open-water oil spill, current response equipment is still inadequate under less than ideal conditions. With existing technology, booms and skimmers alone cannot handle a 10-million gallon spill. Improvements in response technology are needed.

o As in many previous incidents, open-water cleanup attempts evolved into a decision to protect sensitive areas by booming while awaiting shore impacts. Final oil recovery strategies then are based on shoreline cleanup. Significant opportunities to skim oil off   the water in Prince William Sound were lost due to the conflicting priorities of protection.

o Use of dispersants and skimmers in cold-water oil spill responses needs further study. The results need to be shared with industry and incorporated into national plans.

o Beach-cleaning techniques for the conditions found in Prince William Sound are labor- and resource-intensive and not efficient. Further lessons from this experience need to be shared with all concerned.

5. On-Scene Communications

o Communications were a problem, given the large area, the mountainous terrain, and the varied armada of vessels involved. A hodgepodge of radioequipment made communications difficult. Voids in radio coverage were aggravated by, distances and geography inherent to Prince William Sound. Exxon is to be commended for putting together, with Coast Guard and DOD assistance, an effective communications system in short order.

6. Public Information

o The Coast Guard did not initially foresee the magnitude of the public and media interest in this incident. Consequently, the small number of public affairs personnel assigned were overwhelmed.   This adversely affected the OSC organization and its ability to carry out operational responsibilities. Future spills must include prompt, aggressive public information support, including both the means to control misinformation and rumors and the means to coordinate with all relevant agencies, in order to assist the OSC and give the public a more accurate picture of the response.

o A public information program is necessary to deal specifically with the collection, cleaning, and rehabilitation of injured wildlife.

It still is much too early to know the full extent of the environmental, economic, and health effects of the oil that spilled from the Exxon Valdez into the waters off Alaska's south-central coast. The discussion below, which describes these effects, should be regarded as preliminary.

A. ENVIRONMENTAL EFFECTS

1. Background

The severity of oil spill effects on the environment varies greatly, depending on the conditions of the spill. The type and amount of oil involved, its degree of weathering, geographic location, seasonal timing, types of habitat affected, sensitivity of the affected organism's life stage, and adequacy of the response all influence the severity of environmental effects (see Figure 14). Many of the conditions present during the Exxon Valdez spill increased, rather than diminished, the severity of its impacts relative to other large spills. The spill occurred at a high latitude in a semi-enclosed body of water at the beginning of spring. The 10.1 million gallons of oil spilled from the *Exxon Valdez* are known to have oiled over 350 miles of shoreline in Prince William Sound alone. The figure will increase as other affected areas are surveyed.

FIGURE 14

Representation Of Oil Behavior In Prince William Sound

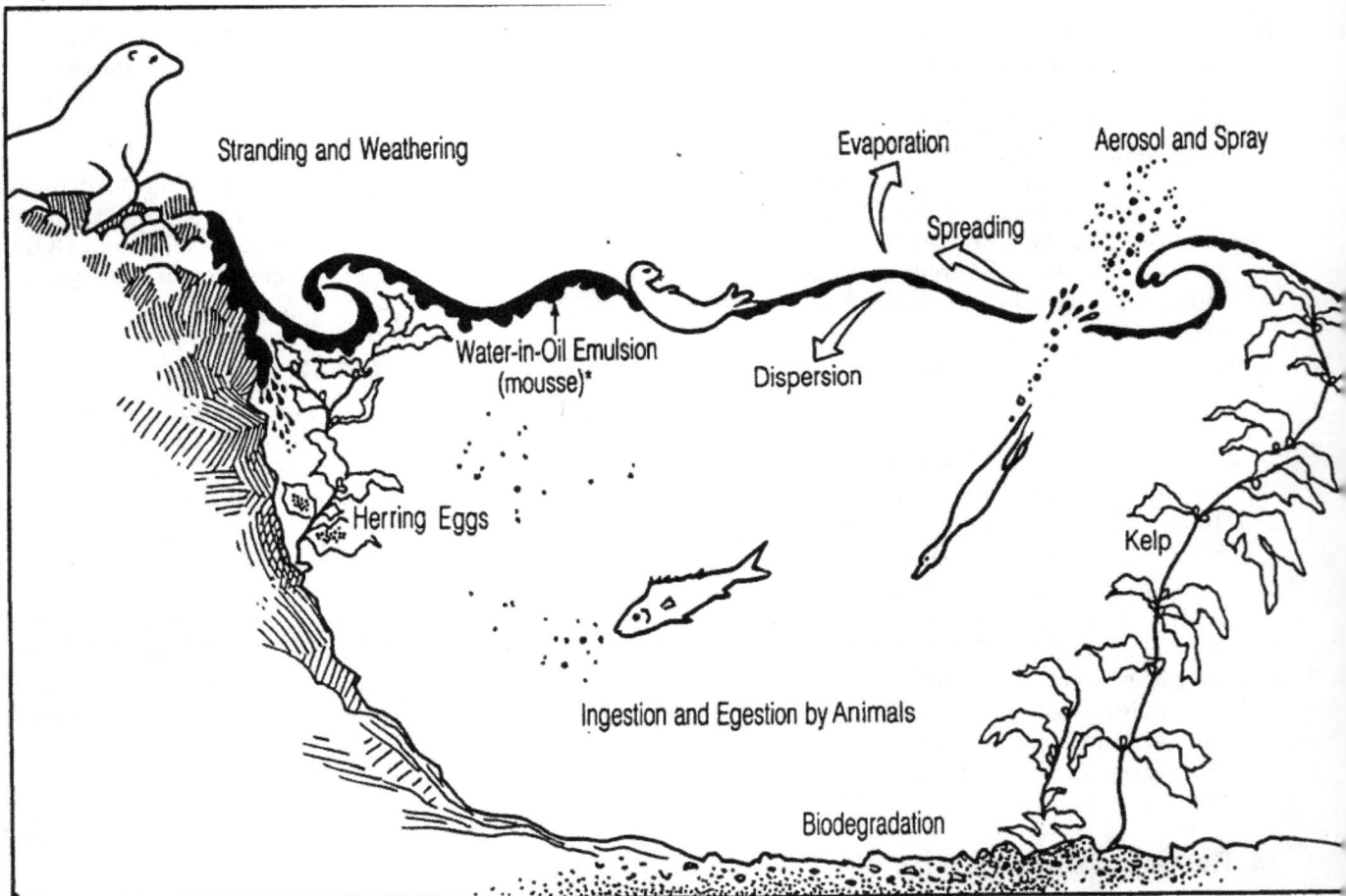

*Mousse is the name given to the thick emulsion of water and oil (50-60% water) caused by the wind and waves.

FIGURE 15
Environmentally Sensitive Areas

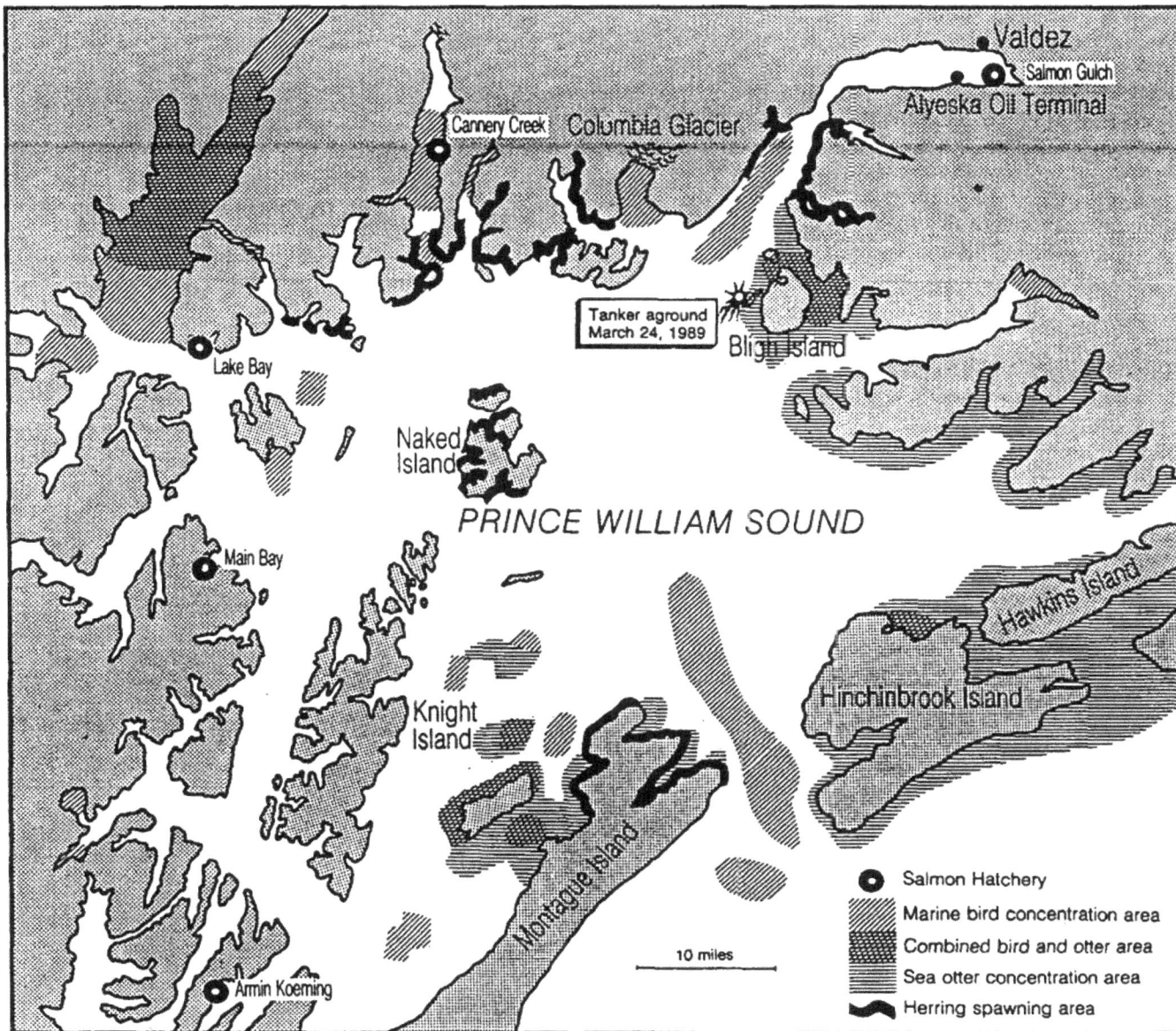

In contrast, only 240 miles of coastline were affected by the *Amoco Cadiz* oil spill in 1978. The *Amoco Cadiz* released 68 million gallons of oil when it broke up on the rocks in stormy seas off France's Brittany coast. Most of the elements of Brittany's temperate zone environment largely recovered within three to eight years from the effects of the oil spill and ensuing clean-up operations. The habitats of the south-central Alaskan coast generally are more vulnerable to spilled oil than those of more temperate climates because the lower temperatures

and resulting slower rates of physical weathering and biodegradation allow the oil to persist. This persistence provides the potential for long-term exposures and sub-lethal chronic effects, as well as short-term exposures and acute effects. In addition, the remoteness of the affected Alaskan area and the physical features of its coastline make cleanup more difficult than it was in Brittany. Great care must be taken in the *Exxon Valdez* cleanup to minimize harm to sensitive environments.

Prince William Sound, the site of the Exxon *Valdez* spill, is one of the largest tidal estuarine systems on the North American continent. In terms of water surface alone, it is about as large as Chesapeake Bay. Its many islands, bays, and fjords give it a shoreline totalling more than 2,000 miles, nearly one-quarter of Chesapeake Bay's total shoreline. Prince William Sound is within the boundaries of the Chugach National Forest. The western half of the sound, the area most affected by the oil spill, is within the Nellie Juan-College Fjord Wilderness study area. This area is highly sensitive environmentally.

Patches of oil or oil-and-water emulsion (mousse) now have moved with the prevailing winds and currents in a southwesterly direction more than 250 miles from the accident site on Bligh Reef (see Figure 16). The oil has moved out of Prince William Sound into the Gulf of Alaska and along the Kenai Peninsula and the Kenai Fjords National Park to the islands of Lower Cook Inlet and the Kodiak Archipelago. There is no evidence to date that large quantities of oil have entered the water column or sunk to the bottom in Prince William Sound.

FIGURE 16

Leading Edge Of Oil Spill (through April 23)

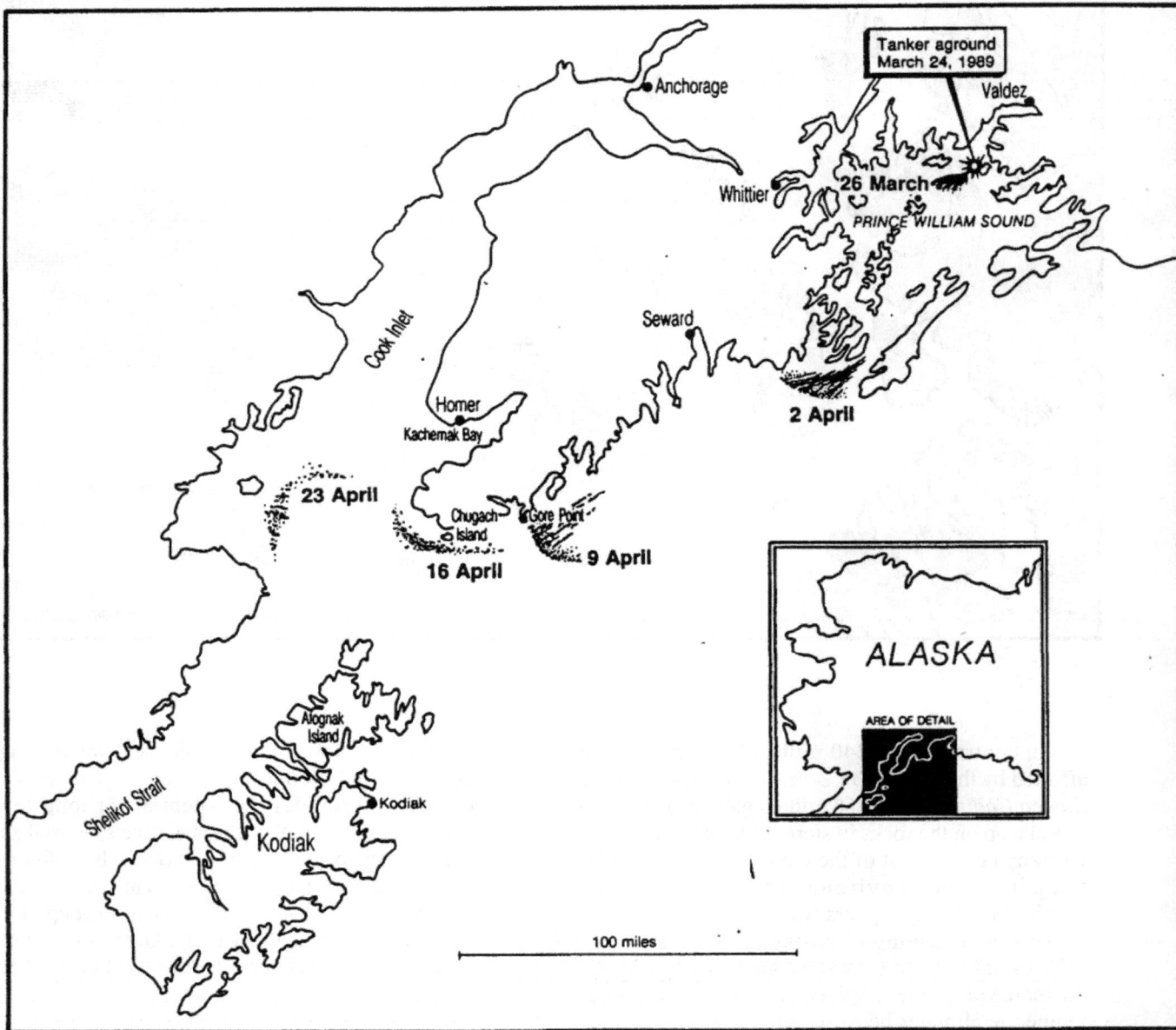

Much of this entire area was largely pristine until the Exxon *Valdez* incident. It is an area of great natural beauty, and its rich natural resources form the basis for major commercial fisheries for pink and chum salmon and Pacific herring. There are smaller fisheries for halibut, sablefish, king,. Tanner and Dungeness crabs, and shrimp. The Chugach National Forest in Prince William Sound and Kenai Fjords National Park are relatively accessible by air and boat from Anchorage, the major population center in Alaska, making the area a favorite location for recreational users. The sound is the major food source for the Alaskan Native villages on its shore.

2. Effects On Birds And Marine Mammals

Immediate spill effects were most visible on marine birds and sea otters.   These effects are becoming much less severe as the oil breaks up into smaller patches and,   finally, into weathered tar balls.

The bird population of Prince William Sound and the Kenai/Kodiak area is diverse and abundant. The Fish and Wildlife Service (FWS) counted more than 91,000 waterbirds (mostly diving ducks, grebes, and loons) in the sound immediately after the spill. About half of these birds were in or near areas affected by floating oil. As the spring migration gets underway, large numbers of waterfowl and shorebirds that stop to feed in the Prince William Sound area potentially could be exposed to the spilled oil. Many of these birds may be affected either directly by oil or indirectly through the loss of food sources.

As the oil moves along the Kenai Peninsula and the Kodiak Archipelago, it will continue to affect shorebirds and waterfowl.   The severity of the impact will depend on the amount of oil that reaches these areas, its degree of weathering and emulsification, and how long it persists near the seabird colonies. Seabirds are just beginning to occupy colonies for this year's breeding season. The success of this breeding season also could be diminished because of habitat loss, loss of food resources, and mortality of chicks and eggs. Oil transferred from the feathers of brooding birds is toxic to embryos within the eggs.   For the reasons discussed above and because of the difficulty in recovering bodies, the 4,463 dead birds collected do not represent the full toll.

Twenty-three species of marine mammals live in the sound and the Gulf of Alaska either year-round or during the summer.   These mammals include gray, humpback, and killer whales, various porpoises and dolphins, harbor seals, sea lions, and sea otters.   Of these animals, the sea otters are by far the most sensitive and vulnerable to spilled oil. Because they are dependent upon fur for insulation, they die of hypothermia and stress when it comes in contact with oil. .Fumes from the floating oil also may have contributed to their deaths. As many as 2,500 of Prince William Sound's estimated pre-spill population of 8,000 to 10,000 sea otters are in the western portions of the sound where they may be exposed to oil from the Exxon   *Vaidez*. The number of dead, currently at 479, is not regarded as an accurate measure of the spill's impact on sea otters because of the difficulty in recovering their bodies. No estimates of total mortality yet have been made.

Other sea otter populations potentially at risk as the oil moves through the Gulf of Alaska off the Kenai Peninsula and the Kodiak Archipelago are the estimated 2,500 to 3,500 otters along the peninsula and the estimated 4,000 to 6,000 around Kodiak and other nearby islands. No other marine mammal (e.g., dolphin, seal, or whale) mortality yet has been attributed to the oil spill, but harbor seals will start pupping in May. There is concern that oil remaining in harbor seal pupping areas could injure or kill the pups. Priority is being given to cleanup of these areas, but the work must proceed cautiously in order to minimize stress on the pregnant females at this critical time: Terrestrial animals, such as river otters, mink, bald eagles, bear, and deer, that utilize intertidal areas, also may be affected through scavenging of oiled carcasses on the beaches or browsing on oiled kelp.

3. Effects On Fisheries
And Other Marine Resources

Oil can affect microscopic plants and animals (phytoplankton and zooplankton) adversely. The latter include the floating eggs and larvae of fish that form the base of the marine food chain. In the open waters of the sound and gulf, this impact probably will be short-lived and local because of the quick replacement of plankton by the same organisms from unaffected areas. For some species, however, mortality of planktonic eggs and larvae may be reflected in long-term population effects. Intertidal animals such as barnacles and mussels, which live in a highly variable and stressful environment, have little or no mobility. Oil in many intertidal areas within Prince William Sound and elsewhere will result in severe mortality among these animals. Recovery of their populations may take several years.

As the oil from the *Exxon Valdez* moves into the deeply indented coast by means of tidal and wind action, it will affect increasingly sensitive environments. Higher-risk, lower-energy environments are located deeper in fjords and bays. In high-energy environments, such as the headlands along the Kenai Peninsula, wave action tends to remove what oil is stranded rather quickly. In low-energy environments, such as shallow bays and marshes, oil may remain for years with only slow chemical and biological processes to degrade it. The stranded oil will serve as a reservoir for the chronic input of oil into the subtidal sediments, where it may affect bottom dwelling (benthic) organisms over the long term. The potential exists for oil released in the *Exxon Valdez* spill to persist in 'and on parts of this coastline for many years.

Long-term effects to the area's rich biota may result from food chain and habitat disruption as well as from decreased survivability and reproductive capability of animals directly exposed to oil. Determining these impacts will require study of the species of concern throughout their life cycle or longer. For example, pink salmon have the shortest life cycle among the five different salmon found in the area. These salmon return to spawn two years after their eggs are laid. Prince William Sound alone accounts for 50 percent of Alaska's total commercial harvest of the species.

A series of state and private hatcheries, two of which are the world's largest, support the pink salmon fishery. Hatchery-raised fry normally are released in early April and spend up to three months feeding and growing in the shallow, near-shore areas of the sound before migrating into the Gulf of Alaska. The fate of this year's fry, estimated to exceed 650 million, is a cause for concern. The fry may be killed by hydrocarbons in their nursery areas (they are sensitive to very low concentrations in the water column). Their growth rate may be slower this year due to stress from hydrocarbons or a decrease in the amount of available food. Because smaller fish are more susceptible to predation, fewer adult fish may return in 1991.

Another economically significant long-term effect could be the possible loss of this year's young herring from the affected areas. Pacific herring are second in importance only to salmon among the fishery resources of the area. Their roe (eggs) provide one of the state's most valuable fish products per unit of weight. The herring and roe fishery in Prince William Sound has been closed this year, and restrictions have been placed on the herring fishery off Kodiak because of the spill. Herring are

spring (April-May) intertidal and subtidal spawners. They do not spawn until they are at least three years old and return each year thereafter during their life span to spawn in their natal areas.

Herring eggs can cover many miles of the intertidal zone. They are both vulnerable and sensitive to oil. The eggs may be smothered and die outright, or oil may cause developmental abnormalities in the growing embryos. The persistence of stranded oil in herring spawning areas may affect not just the 1989-year class but also subsequent-year classes. This impact can be determined best by examining the spawning adults at areas of impact in 1992, 1993, and 1994 for the percentage of the population recruited from spawn in 1989 through 1991.

4. Federal And State Action To Address Environmental Impact

Section 311(f) of the Clean Water Act (CWA) authorizes the President and state officials to act on behalf of the public as Trustees for natural resources seeking recovery from Exxon for the costs of restoring, rehabilitating, or acquiring the equivalent of the injured resources. The State of Alaska and the U.S. Departments of Agriculture, Commerce, and the Interior have primary trust resources affected by the spill.

Representatives of the Trustee agencies are working together in Alaska to develop a plan for assessing the short- and long-term effects of the spill on their Trust resources, the extent of the injury, the resulting economic damages, and the probable cost of restoration. This plan is scheduled tentatively for completion in early June.

Exxon has agreed to make up to $15 million available initially to the Trustees to fund these damage assessments, but Trustees believe that additional funding will be needed. Because some of the potentially affected resources, such as salmon, have two- to seven-year life cycles, estimating the extent of their injury will require long-term studies. It may be difficult therefore to assess fully all injuries and economic damages for a number of years. Where necessary and possible, restoration will require even more time.

The need for information about the environmental aspects of the *Exxon Valdez* spill extends beyond what will be learned as a result of assessing injury and economic damages to natural resources. The President has asked Environmental Protection Agency (EPA) Administrator William Reilly to coordinate the long-range planning to restore the environment of the sound. The expertise of leading governmental and private scientists and oil spill experts will be used in this work. To the extent feasible, existing coordinating groups such as the National Ocean Pollution Policy Board, an inter-agency group established by Congress for the coordination of marine pollution research development and monitoring, will be included. Other existing federal scientific advisory and coordinating bodies will participate if possible.

5. Early Lessons Learned/ Recommendations

o The Departments of Agriculture, Commerce, and the Interior, as federal Trustees for the affected natural resources, should work closely with the state trustee agency, the Alaska Department of Fish and Game, to plan and implement natural resource damage assessments as quickly as possible. In doing so, they should coordinate their activities with response authorities to avoid interfering with the cleanup.

o The Trustees, together with EPA, should work in coordination with the parties assessing long-term environmental effects to avoid duplication and develop the best possible scientific basis for restoration of Prince William Sound and other affected areas.

o Where applicable, results of past studies should be used. New research should be used to confirm earlier preliminary findings *or* to fill gaps.

o In future spills, damage assessment and restoration should begin immediately and funding options should be identified quickly.

o To facilitate response to future incidents, federal Trustee agencies should develop an automatic mechanism to resolve in advance such issues as identification of lead Trustee, management of assessment funding, delineation of restoration responsibilities, and the allocation of restored funds recovered from joint claims.

o Federal agency damage assessment capabilities should be strengthened so that a small cadre of trained and experienced personnel will be able to go immediately to the scene of major spills in the future.

o States that have not yet done so should be encouraged to designate Trustee agencies as provided under §107(f)(2)(B) of the Comprehensive Environmental Response, Compensation, and Liability Act, and §311(f)(S) of the CWA.

o Wildlife rescue efforts need to be implemented immediately after a spill is reported. In addition, research procedures should be established quickly to allow data collection required to develop improved rescue efforts in the future.

B. ENERGY EFFECTS

1. Importance Of Alaska North Slope Oil

Alaskan North Slope (ANS) crude oil is produced in the Prudhoe Bay area of Alaska in the northern part of the state. During 1988, ANS crude oil production made up about one-quarter of U.S. crude production and about 12 percent of US. petroleum consumption. Approximately two million barrels of ANS oil per day is transported by a 48-inch pipeline 800 miles to the Port of Valdez.

All ANS crude oil remains within the United States and US. territories. During 1988, about 70 percent of ANS crude oil was transported to the west coast, and slightly over 15 percent was transported to the Gulf coast. The remainder went to the east coast, the midwest, and the U.S. territories.

2. Market Impact Analysis

The Department of Energy(DOE) intensified its monitoring of available energy supplies and fuel prices immediately after the *Exxon Valdez* spill interrupted Alaskan crude oil shipments. In the three weeks following the spill, gasoline prices rose, but this increase was only temporary. Los Angeles spot gasoline prices rose by 50 cents to $1.18 per gallon at their peak on March 31. Nationally, unleaded regular gasoline prices increased, on average, about 10 cents per gallon at both wholesale and retail levels. Complete data for a thorough analysis of market responses to this event are not yet available. Nonetheless, some observations and preliminary conclusions can be made.

Over the 13-day period before the *Exxon Val-dez* was refloated, the reduction of Alaskan crude oil production was approximately 13 million barrels (see Figure 17, which shows pipeline throughput quantities that are roughly equivalent to production quantities). This amount is small by national standards. It represents the equivalent of about 17 hours of total national petroleum consumption and less than two percent of total annual production from the ANS which eventually leaves the Port of Valdez. Approximately 10 million of these 13 million barrels, however, **had** been destined for west coast refiners, and that region suffered a temporary, disproportionate increase in retail gasoline prices. The interruption of Alaskan crude oil also may have contributed to price increases by creating serious concern regarding future supply curtailments in product oil markets.

FIGURE 17
Alaskan Oil Pipeline Throughput

Millions of Gallons Per Day

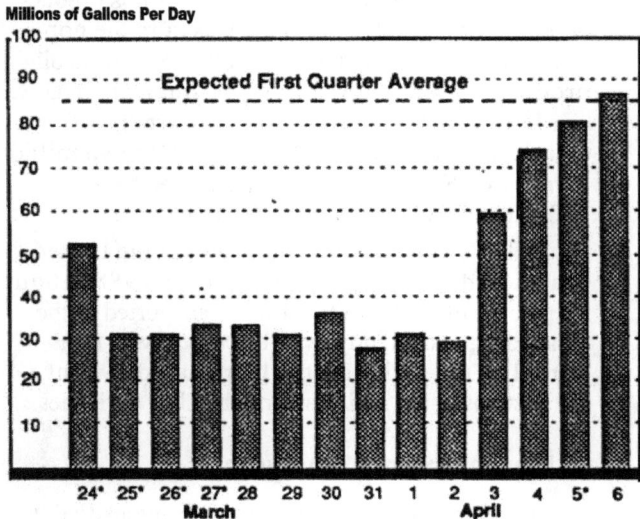

Source: U.S. Department of Energy, Energy Information Administration, 1989.
* No tanker transport on these days.

In the two weeks following the spill, crude oil from other producing regions offset west coast crude oil losses by drawdown of refinery crude oil stocks, by drawdown of product stocks, and by product imports. Three weeks after the spill, other crude supplies and crude oil accumulated at the Valdez terminal nearly had compensated for the earlier supply loss. The temporary reduction in production should have no lasting or permanent effect on gasoline prices.

Observed west coast retail gasoline price increases, which began in February long before the spill, can be attributed to several causes. Crude oil prices (West Texas Intermediate) rose from a two-year low of $12.58 per barrel in early October of

1988 to slightly. above $20 by **the** day before the *Exxon Valdez* incident. This price increase of $7.50 per barrel translates into an equivalent increase in product prices of 18 cents per gallon.

Both average U.S. wholesale and retail prices of gasoline, however, lagged behind the crude oil price increase by about four months. Wholesale price increases due to the crude oil increase had begun to rise before the *Exxon Valdez* incident. Yet average U.S. gasoline retail prices for the nation as a whole had remained essentially unchanged over the period from early October 1988 to mid-March of 1989. Other factors also may have contributed to the rise in west coast gasoline prices since February. These factors include: strong seasonal gasoline demand, lower refinery gasoline inventories due to routine seasonal refinery maintenance, and new gasoline specifications that reduce vapor emissions but increase refining costs.

Fortunately, shortages have not occurred. Prices in California have fallen from their reported peaks. In the week ending April 17, Los Angeles spot gasoline prices fell from $1.18 per gallon to $0.79 per gallon. Market forces have operated to provide California markets with energy following the brief interruption of Alaskan crude oil.

3. Early Lessons Learned/ Recommendations.

o A variety of factors contributed to product price increases observed following the *Exxon Valdez* spill. Among these factors may have been concerns regarding future supply curtailments in product markets. Overall, however, market forces appeared to have operated efficiently to meet energy demands.

o Despite the tanker transportation safety record out of Valdez, an incident like the *Exxon Valdez* accident can crystallize public opinion against the petroleum industry almost instantaneously.

o Conflicts need to be resolved on issues such as the use of dispersants, the risk of fire, and the state of readiness.

o Workable oil spill contingency plans and sufficiently trained response personnel, along with policies and practices to police the industry workforce, must be in place. Industry must operate at all times with a clear recognition of the importance of environmental safeguards and adequate responsiveness. Most important, environmental protection should not be just another regulatory burden, but the watchword of every aspect of operations.

C. EFFECTS ON THE ALASKAN ECONOMY

The natural resources of the areas affected by the *Exxon Valdez* spill are important to Alaska's local and statewide economy. While it is too early to know the full extent of the economic consequences of the spill, the local' and state economies are likely to suffer economic losses in the following categories.

1. Commercial Fisheries

Prince William Sound possesses rich commercial fisheries for Pacific herring and salmon, along with smaller halibut, sablefish, crab, and shrimp fisheries. These fisheries are used on a permit basis by commercial fishermen from as far away as Anchorage and Seattle. The Alaskan fishing ports nearest the sound are Cordova, Seward, Homer, and Kodiak. Cordova, probably the most affected of the fishing ports, is the third largest in Alaska and the ninth largest in the United States in terms of the dollar value of commercial fishery landings. Kodiak is the largest Alaskan fishing port and the second largest in the United States.

Together, these two ports had commercial fishery landings of all species valued at $174 million in 1957. This catch represented over 18 percent of the total for Alaska and nearly six percent for the United States in that year. An estimated one-third of Alaska's nearly 12,000 full- or part-time fishermen in 1987 worked in the area now affected by the spill.

Prince William Sound's herring and herring roe fishery (valued at $14 million in 1988) usually opens in early April. Out of concern for additional harm to the stocks and possible contamination of the product, the Alaska Department of Fish and Game closed the herring fishery after the spill and recently restricted part of the herring fishery off Kodiak. It is unknown at this time whether or not the $33-million pink salmon fishery in the sound, which reaches its peak in July and August, will be closed or restricted for similar reasons.

Closings or restrictions *will* harm not only fishermen but also the area's important fish processing industry. This industry employs an estimated 3,000 to 4,000 people annually. The State of Alaska, with technical assistance from the Food and Drug Administration (FDA), is taking precautions to assure that oil-tainted fish products do not reach the market. The state is assigning 40 extra inspectors to the processing plants serving the affected area and will continue to close fisheries, if necessary, to protect the public (see Section D below). If, despite these measures, consumers avoid Alaskan fish products, the national prices for those products may be depressed temporarily.

With the spill still spreading, the full economic impact on commercial fishermen is unknown. The immediate economic losses of many local fishermen are being mitigated by their employment in Exxon's clean-up efforts. Fishermen in the affected area remain deeply concerned not only about their long-term economic prospects, but also about possible changes in their way of life.

2. Recreation

Recreation and tourism have been increasing rapidly in Prince William Sound over the last 10 years. In the late 1970's, cruise ships did not visit the sound, but, by 1987,. ship visits had reached 88 per season. In the same year, an estimated 1.8 million people visited Prince William Sound for recreation purposes. Much of the recreation and tourism in Prince William Sound and the Kenai Fjords National Park is related to the outstanding scenic beauty of the area and its pristine wilderness character. The Alaska National Interest Lands Conservation Act (ANILCA) created a 2.3-million acre wilderness study area in Prince William Sound.

The oil spill can be expected to affect tourism and recreation in the affected region of Alaska at least through the approaching summer season. The tourist industry already is reporting higher than normal cancellations on bookings for this summer. The magnitude and duration of these adverse consequences will depend in part on the speed and effectiveness of the cleanup and in part on the public's perception of its effectiveness in restoring the wildlife and scenic areas to their pre-spill condition. The spill. is not expected to have a major detrimental impact on travel and tourism in the rest of Alaska.

3. Native Villages

Native villages such as Tatitlek and Chenega on the shores of Prince William Sound depend on the animals, birds, fish, and plants of the sound and surrounding lands for their food. The existence of their traditional culture depends on the continuation of this subsistence economy. Losses or reductions in the availability of wild food sources cannot be measured adequately in dollars.   Although the Natives are gaining some employment opportunities from Exxon's clean-up efforts, they remain deeply concerned about the long-term effects of the spill on their subsistence culture. There also is concern that beached oil and its cleanup may either destroy cultural resources or affect the ability of archaeologists to carbon date early sites. No single mechanism is in place at this time through which Alaska Natives can provide inputs on their particular concerns, or receive assistance for their claims and subsistence needs.

4. Timber

Neither the oil spill nor its cleanup is expected to affect timber harvesting on national forest lands around Prince William Sound and on the Kenai Peninsula, and economic losses are unlikely. Some delay, however, in harvesting on Native Corporation and national forest lands on Montague Island may result from the spill. The U.S. Forest Service has had to extend the review period for the draft Environmental Impacts Statement because of oil spill response activities.

5. Early Lessons Learned/ Recommendations

o The Department of the Interior, working with the State of Alaska and local Native leadership, should assist individual Alaska Natives and Native organizations in providing input into clean-up planning and filing claims for economic losses. This assistance should include the emergency provision of subsistence needs wherever required as a direct result of the spill.

o There is no existing legislation that allows immediate aid to the local population affected by the spill. In this incident, Exxon mitigated some of the economic losses to fishermen, Alaskan Natives, and other Alaskans through employment in the clean-up effort. Had there not been a responsible party who willingly assumed this financial burden, there would have been no immediate financial relief available to the affected population.

D. HEALTH EFFECTS

Potential human health impacts from the *Exxon Valdez* oil spill include those associated with exposure to contaminants either directly or through the food chain; stress associated with loss of lifestyle and possible economic impacts; and hazards workers may encounter during clean-up operations.

1. Food Safety

It is important to keep dead fish and mammals out of the food chain and allow harvesting of seafood only when its safety has been assured. Currently, organoleptic (sensory) testing is being used to determine whether fish and shellfish should be consumed.   This testing method can be used to determine freshness and presence of volatile oil components. The FDA is now training state sanitarians in organoleptic techniques.

To ensure food safety properly over the long term, however, a system involving both organoleptic and chemical analytic testing needs to be developed. FDA and other agencies and experts are obtaining information on the characteristics of crude oil, accumulation and dissipation of oil components in fish and shellfish, analytical methodology, organoleptic detection in fish and shellfish, and toxicity data relative to long-term consumption of oil components. Information developed from these sources will be evaluated, particularly in terms of the type of food safety program needed.

2. Mental Health

With the assistance of a contracted disaster psychologist, the State of Alaska is now assessing the nature and extent of mental health problems resulting from the spill among emergency workers, fishermen, and others and present capabilities to address these problems. The state plans to enhance the service capabilities of its system based on this assessment.   The state will seek payment from Exxon for short-term supplemental assistance to the mental health system.   U.S. Health and Human Services agencies such as the Alcohol, Drug Abuse, and Mental Health Administration (ADAMHA) are prepared to offer technical and other support.

3. Occupational Health And Safety

Since the spill, Exxon has employed boat owners to assist in skimming and booming operations. Exxon estimates the company will hire between 2,000 and 5,000 people to clean up oil-covered shoreline. This cleanup is expected to continue through the summer. The potential for worker injury and other problems exists.

Workers have expressed concern to both state and federal authorities about safety and health risks in performing clean-up operations. These risks include: inhalation exposure to volatile, and dermal exposure to non-volatile, components of crude oil; exposure to chemical dispersants; stress from long hours; possible physical injury and hypothermia; and lack of available information about the health effects of materials being used.

Regulations that address these concerns are in place. Alaska's Department of Labor is providing advice to Exxon on personal protective equipment and required training in hazard recognition and prevention. The state also has established a field headquarters in Valdez to improve communications and monitor events. In accordance with the regulations, a field headquarters for the Occupational Safety and Health Administration (OSHA) has been established in Valdez, and Exxon has set up a training program for the clean-up crews. The program covers areas such as proper clothing, hazard recognition, and first aid and injury protocols.

4. Early Lessons Learned/ Recommendations

Short-term

o FDA and the State of Alaska should undertake biological monitoring of potentially affected fish and mammals in the spill area on a continuing basis. Strict guidelines for reporting and comparing analytic data should be determined in collaboration with the Centers for Disease Control (CDC) and the Agency for Toxic Substances and Disease Registry (ATSDR). All relevant information should be disseminated centrally within the State of Alaska.

o The mental health system should be monitored continually by the State of Alaska, supported by ADAMHA, to ensure its adequacy to meet the mental health needs of the population.

o Worker training, appropriate protective equipment, and occupational safety and health surveillance should continue to be emphasized by Exxon and coordinated with OSHA, organized labor, the Alaska State Department of Health, and the National Institute for Occupational Safety and Health.

o The State of Alaska's Health Department has relayed concerns from workers and the community regarding long-term effects of exposure to oil chemicals and clean-up chemicals, and long-term effects of consuming food possibly contaminated by these chemicals. The U.S. Department of Health and Human Services (primarily through ATSDR, CDC, and FDA) should continue to provide any needed technical assistance to the state to assess the long-term health effects of exposure to the spill of crude oil, its degradation products, and dispersants used in the clean-up effort.

VI. LIABILITY AND COMPENSATION

A. DESCRIPTION OF COMPENSATION AND LIABILITY PROVISIONS

Exxon's exposure to liability for the grounding of the *Exxon Valdez* stems from the Clean Water Act (CWA), the Trans-Alaska Pipeline Authorization Act (TAPAA), general maritime law, and Alaska state law.

Although aspects of both §311 of the CWA and TAPAA apply to the *Exxon Valdez* spill, it is unclear to what extent they may both be applicable or how they will interrelate. As indicated, state and general maritime tort laws also apply, but the legal complexities of the incident ultimately may require resolution in the courts.

Under the CWA, an owner of a vessel is liable for both clean-up costs that may be incurred by the federal government and for damage to natural resources under the trusteeship of the federal and state governments. Absent a successful defense, Exxon has approximately a $14.3 million liability cap. If "willful negligence or willful misconduct" can be proven, however, Exxon's liability under the CWA is unlimited. Without proof of willful negligence or misconduct, §311(f)(l) of the CWA limits Exxon's liability to $150 per gross ton of the vessel ($14.3 million). Each natural resource damage claim is prepared by the Trustees: the Department of the Interior, the National Oceanic and Atmospheric Administration (NOAA) of the Department of Commerce, the Department of Agriculture, and the State of Alaska.

Section 311(k) of the CWA authorizes creation of a revolving fund to finance, among other things, the removal of oil and hazardous substances spilled from vessels. The §311(k) Fund has been used since the first day of the spill. It has not been used to finance federal removal costs; rather, it has served as authorized by the CWA to facilitate Exxon's use of federal resources. As of April 14, $13 million had been spent or obligated from this fund in connection with the use of those resources. Through that date, Exxon had reimbursed the §311(k) Fund a total of nearly $10 million.

The Trans-Alaska Pipeline Liability Fund (the TAP Fund) is available to pay damage claims resulting from spills of oil that has been transported through the Trans-Alaska Pipeline and loaded on a vessel to be carried to another U.S. port. The TAP Fund was created as part of the statute in which Congress authorized the construction and operation of the Trans-Alaska Pipeline in 1974. Money charged on each barrel of pipeline oil loaded on a vessel financed the fund. The amount of money currently in the fund is approximately $250 million.

Under the statute and the regulations promulgated by the Secretary of the Interior to govern the TAP Fund's administration, the owner and operator of the vessel from which the oil is spilled are each strictly liable for the first $14 million in damages resulting from the oil spill. Owners and operators of vessels are required to furnish proof of their financial responsibility for this liability. Owners must present the proof before their tankers may be loaded with North Slope crude oil at the pipeline's terminal in Valdez.

Accordingly, under the statute and regulations Exxon is liable for the first $14 million in claims. Once that amount is paid, the industry-supported TAP Fund provides an additional $86 million for claims. The TAP Fund has the right to recover money it has paid on claims if it can be shown that Exxon was negligent or that the *Exxon Valdez* was unseaworthy. Under those circumstances, Exxon's total liability under the TAP Fund could reach $100 million. TAPAA specifically does not preclude recovery under state or other federal law.

TAPAA covers all damages, including either public or private clean-up costs sustained by any person or entity. It also covers claims by Canadian residents. If total claims exceed $100 million, each claim for TAPAA funds is reduced proportionately. In such a case, the TAP Fund considers claims after the response is completed and total claims are known. The TAP Fund is not designed to support ongoing response actions.

When spills of oil from tankers carrying North Slope crude oil occur, the owner and operator designate a single contact person who coordinates with the TAP Fund the resolution of claims arising from the spill. For spills exceeding a projected $14 million in damages, the TAP Fund advertises availability of the Fund and the person to whom claims should be directed.

Exxon has designated a contact person to coordinate actions with the TAP Fund and is processing claims. The TAP Fund will publish an advertisement in the very near future specifying that claims should be presented to Exxon. The advertisement will set forth addresses and telephone numbers for presentation of claims.

The §31 l(k) Fund had only $6.7 million available when the spill occurred. The fund has an authorized ceiling of $35 million, but its highest end-of-year balance was $24.8 million in 1985. Since fiscal year 1986, the balance has been less than $12 million. In part, this situation was due to the anticipated passage of proposed comprehensive oil spill legislation.

There is no corresponding fund to finance natural resource damage assessments. Money for natural resource damage assessments can be placed in the §311 (k) Fund by the discharger for the use of Trustees. Initially, the natural resource Trustees funded a preliminary assessment by diverting funds from other programs. The federal and state natural resource trustees, working in conjunction with the Department of Justice, approached Exxon and were able to obtain a $15 million commitment for initial funds to begin natural resource damage assessments.

If Exxon had not voluntarily assumed financial and clean-up responsibility for the spill, the §311(k) Fund probably would have been rapidly expended. Furthermore, there might not have been adequate money available for resource damage assessment and restoration. Legislation under consideration would create a new fee-based fund with a significantly higher balance than in the §311(k) Fund and liability regime. This new fund would be available not only for cleanup, but also for resource damage assessment and third party damage recovery.

B. EARLY LESSONS LEARNED/ RECOMMENDATIONS

o Congress should enact comprehensive oil spill liability and compensation provisions along the lines of the legislation proposed by the Administration. It should include implementation of: (1) the 1984 Protocols to the International Convention on Civil Liability for Oil Pollution Damage, 1969 and (2) the International Convention on the Establishment of an International Fund for Compensation for Oil Pollution Damage, 1971. Enactment of such legislation would make available means in all cases to address adequately oil tanker spills that could extensively damage our coast.

o The relationship between liability requirements under the Clean Water Act and other statutes is undefined and could result in costly and extensive litigation. 130th total liability and necessary compensation in the case of the *Exxon Valdez* spill remain undetermined. Cleanup, natural resource restoration, and third-party damages will be enormous. Had Exxon not made vast sums of money available rather quickly, or had the discharger been unreachable, foreign, or less solvent, the patchwork of existing federal and state law applicable to a pollution incident of this magnitude would have been inadequate.

o Laws and regulations on handling the §311(k) Fund need to be analyzed. Money to finance natural resource damage assessments currently is being placed in the pollution fund by Exxon to be disbursed to natural resource Trustees. The Coast Guard is responsible for receiving and disbursing that money to the Trustees. The responsibility of the Coast Guard, vis-a-vis the Trustees, however, is undefined and requires further review.

o The Clean Water Act needs analysis in relation to other existing law to determine whether other issues associated with major marine disasters resulting in large spills are addressed adequately. In particular, the need for additional criminal sanctions, civil penalties, and judicial and administrative order authority should be examined.

VII. GENERAL LESSONS LEARNED/RECOMMENDATIONS

The *Exxon Valdez* oil spill severely tested this country's existing oil spill preparedness and response capabilities. It revealed shortcomings that require immediate attention. While definitive conclusions about many aspects of the incident remain premature, a number of important lessons and recommendations have emerged in the course of developing this report. Significant actions to prevent or mitigate the impacts of similar tragedies already are underway. Certain steps that warrant careful attention in the near future can be identified.

Wxitw$xlex$Epvieh}$L ezi$F iir$Mrmmexih

o Several investigations, independent of this NRT report, are underway to investigate the cause of the spill. The National Transportation Safety Board, Coast Guard, State of Alaska, and other authorities concurrently are looking into different aspects of the spill. These investigations will create a more complete picture of its causes and suggest strategies for strengthening prevention.

o The President has asked the Environmental Protection Agency Administrator to coordinate long-range planning to restore the environment of Prince William Sound. The Administrator is establishing a task force to this end. This task force will bring together the expertise of leading government and private scientists and the interest of the public in an advisory capacity. This work will yield important information particular to Prince William Sound and will advance our general state of knowledge concerning both the long-range environmental impacts of oil spills and ways of ameliorating their impacts.

o Comprehensive oil spill liability and compensation legislation is pending. Such legislation is a necessary prerequisite to ratifying the 1984 Protocols to the 1969 Civil Liability and 1971 Fund Conventions. These steps will address a number of troubling questions about the extent of corporate liability for oil spillers world-wide.

o A number of federal agencies are coordinating closely with the State of Alaska to undertake natural resource damage assessments to learn about the short- and long-term environmental. effects of the spill, resulting economic damages, and the probable cost of recovery. These studies will pave the way for further steps to reduce adverse environmental and economic effects.

o Under the leadership of the Coast Guard, the National Response Team (NRT) is conducting a six-month study of contingency planning. Preparedness must be improved. The study will examine the use of worst-case scenarios to ensure realistic planning and the need for thorough testing of plans with exercises. The study will address questions associated with the adequacy of equipment and personnel for an effective response, and the importance of well-defined organizational responsibilities.

o A panel of scientists has been established to review the feasibility of using bioremediation techniques to clean up oil spills and to design demonstration projects to evaluate these techniques for use after oil spills.

Psrklxivq $K sepw

o Improved response coordination between federal, state, and local authorities can facilitate rapid clean-up actions. Complications associated with the *Exxon Valdez* oil spill response highlight the importance of smooth coordination. A better way of incorporating the concerns of states into the National Response System (NRS) should be developed. The NRT also will undertake a program to improve understanding of the National Response System among top local, state, and federal officials. Improved understanding is an essential basis for smooth coordination.

o The NRT must initiate a study of ways to improve the National Contingency Plan (NCP). The NCP has been effective in minimizing environmental and health impacts from accidents for over 18 years, but changes that ensure the optimum response structure for releases or spills of national significance require study.

o Attention should be paid to Native Alaskans whose livelihoods have been affected by the spill and whose cultural practices may have been threatened. Short-term and long-term steps to mitigate economic and mental health impacts may be required.   The Department of Interior, working with the State of Alaska and local Native leadership, will work together to provide assistance.

o Steps must be taken to improve overall planning for, and care of, wildlife affected by oil spills. Contingency planning should be expanded to prepare properly for wildlife impacts. Signifi-cant actions to prevent or mitigate similar tragedies already are underway.

o The *Exxon Valdez* incident emphasized the need for greatly improved public and private research and development capabilities. Current response equipment is still inadequate in less than ideal conditions. Better mechanical, chemical, and biological strategies for cleanup are needed. The incident revealed how little we know about cold-water oil spill responses.   The oil spill showed the need to develop beach-cleaning techniques that are more effective and less labor-intensive.

o Possibilities to strengthen 'existing international ties should be explored.   Working through federal agencies involved in established organi-zations, agreements addressing available equip-ment and personnel for spills of this magnitude should be pursued. Better coordination of joint efforts to transport and stage equipment from other countries would enhance response capabil-ities.   Joint research and development and information sharing agreements between nations also warrant consideration.

APPENDIX A. CHRONOLOGY

(Derived from U.S. Coast Guard Pollution Reports. All times are local
Hours of daylight are roughly from 0600 to 2200.)

MARCH 24,1989 (FIRST DAY)

0028 Marine Safety Office (MSO) Valdez receives notification from *Exxon Valdez* reporting that the vessel was hard aground on Bligh Reef. The vessel was loaded with 1,264,155 barrels (53,094,510 gallons) of North Slope crude oil. While maneuvering to avoid glacial ice, the vessel left the tanker lanes and struck bottom on a 30-foot charted shoal on Bligh Reef. The vessel's draft fore and aft was 56 feet. Approximately 510,000 gallons of crude oil released. Initial assessment of pollution extent and shoreline impact could not be made with any degree of accuracy due to darkness.

0030 Captain of the Port (COTP) at MSO Valdez closes Port of Valdez to all traffic. The tug *Stalwart* is dispatched from Alyeska Marine Terminal to assist Exxon *Valdez*.

0100 Pilot boat is made available to transport Coast Guard (CG) and Alaska Department of Environmental Conservation (ADEC) pollution personnel to the scene.

0105 Alaska RRT Chairman for this incident is notified.

0148 MSO Valdez contacts Air Station Kodiak requesting helicopter overflight at first light.

0206 Attempt to refloat *Exxon Valdez* at high tideunsuccessful.

0227 Motor vessel (M/V) *Shelikoff* reports oil slick to half mile south of Exxon *Valdez.*

0249 Pacific Area (PACAREA) Strike Team assistance requested.

0323 CG personnel now aboard *Exxon Valdez.* Tanks gauged: about 138,000 barrels (5.8 million gallons) lost from wing tanks 1, 3, 5 starboard, starboard slop tanks, and number 5 center tanks.

0330 Initial response efforts at the Port of Valdez under Alyeska's control are hampered by equipment casualties and holiday personnel shortages. As response personnel arrive at the Alyeska Terminal, however, Alyeska is unable to comply with the response. timeliness provision in its own contingency plan that calls for initial response at the vessel within five hours of first notification.

Alyeska's only containment barge is tied up at Valdez Terminal, stripped for repairs. Barge was not certified by the CG to receive oil, but it could carry recovery bladders. Alaska's state contingency plan requires Alyeska to notify the state when response equipment is taken out of service. Satisfied the barge was seaworthy without repairs, Alyeska had not done so.

Before barge could be used, pollution gear had to be loaded. Crane riggers called at 0330. By this time, CG estimates 5.8 million gallons already discharged from the tanks.

0400 Stability of *Exxon Valdez* is CG's highest priority. The biggest concern is that the vessel might capsize and break up, spilling the entire 53 million gallons of oil. Word is passed to Alyeska to assemble oil transfer (lightering) equipment (six-inch hoses and fenders to transfer oil from *Exxon Valdez* to recovery vessels, bladders or other containers).

0414 Tank vessel (T/V) *Exxon Baton Rouge* contacted and is enroute to initiate oil transfer operations.

0500 CG small boat dispatched to survey the area.

0727 Alyeska Marine Terminal security helicopter aloft for overflight with CG investigator aboard. Analysis of videotape taken by the investigator showed the slick to be 1,000 feet wide by four miles long. Sheen extends in a southerly direction with minimal contact with shoreline.

1115 The Federal Aviation Administration (FAA) imposes temporary flight restriction around *Exxon Valdez*. Rate of discharge from *Exxon Valdez* is slowing. Exxon Baton Rouge arrives at location of grounding and begins to rig fenders for coming alongside to begin oil transfer operation

1140 H-3 helicopter overflight observes extremely heavy oil 20 to 30 feet from the side of the vessel. Calm sea is slowing the movement of the oil.

1145 COTP imposes 500-yard safety zone around grounded vessel. Notice to Mariners broadcast begins.

1200 Regional Response Team (RRT) teleconference commences. Discussion includes use of dispersants and in-situ burning.

1230 Alyeska barge arrives at Bligh Reef, seven hours after the five-hour initial response time required by Alyeska's contingency plan. Oil slick is already 1,000 feet wide and four to five miles long. The barge arrives with two skimmers in tow, two 1,000-gallon bladders, and 8,000 feet of containment boom for a spill of 10 million gallons. The two on-scene skimmers begin recovering oil near the *Exxon Valdez.*

1310 MSO Valdez estimates quantity released is now 250,000 barrels (10.5 million gallons).

15 10 On-Scene Coordinator (OSC) grants permission for dispersant test on leading edge of sheen.

1645 CG 32323 underway with State of Alaska Governor Cowper and assistants on-board.

1700 CG 32323 drops off Governor Cowper and assistants. They board *Exxon Valdez.*

1800 Dispersant trial application is conducted with less than satisfactory results, due to lack of mixing energy. Use of dispersants is deemed inappropriate at this time.

1820 PACAREA Strike Team members arrive in Cordova, Alaska.

2010 *Exxon Baton Rouge* alongside *Exxon Valdez,* port-to-port.

2030 Contract divers arrive on scene.

2154 *Exxon Baton Rouge* made fast alongside *Exxon Valdez.*

2215 First oil transfer hose connected.

2338 Second oil transfer hose connected between the two vessels.

MARCH 25,1989 (SECOND DAY)

0015 Tug *Jeffrey Foss* on scene with 30,000-barrel tank barge to receive recovered oil.

0736 Oil transfer operations begin, with cargo transferred to *Exxon Baton Rouge.*

0745 OSC reports that loss of additional oil has ceased.

0750 Commercial divers complete underwater hull survey showing holes in 11 tanks. *Exxon Valdez* had been grounded from the number two tank aft to the number four tank. The vessel's designer advises CG that ship is not floatable and that a major salvage operation will be required. Meanwhile, oil transfer to the *Exxon Baton Rouge* continues at a rate of 10,000 to 12,000 gallons per hour. Water replaces cargo to maintain ballast.

0830 Alyeska pipeline representative advises that pipeline flow has been slowed to 768,000 barrels per day. At this rate, Alyeska has seven days of storage capacity.

0833 USCGC *Rush* is ordered to put CG personnel aboard T/V *Oriental Crane* and to proceed with all haste to *Exxon Valdez* site and establish a command and control platform. USCGC *Sedge* is ordered to Prince William Sound to assist OSC.

0845 Oil transfer operations suspended as on-scene crews reevaluate the situation.

0930 Ms. Alice Berkner of International Bird Rescue and Research Center arrives in Valdez to set up facilities for treating oiled animals.

0945 On-Scene RRT agency representatives hold a meeting at MSO Valdez with teleconference following at 1110. Dispersant trial application planned utilizing C-130 cargo aircraft. Exxon Shipping Company assumes management of spill and financial responsibility.

1100 *Exxon Valdez* surrounded by containment boom. USCGC *Rush* on scene to assume air traffic control functions.

1145 CG Commander of the Pacific Area (PACAREA) requests AIREYE surveillance from the Atlantic Area Commander.

1200 Second hull survey conducted with video cameras by dive team.

1230 Some oil released as transfer operations begin again. Oil transfer had been delayed due to piping damage.

1330 FAA in Anchorage is mobilizing air traffic control team to set up temporary control tower (seven personnel) at Valdez Airport.

1900 RADM Nelson, Commander Seventeenth CG District, arrives on scene.

1930 Clean-up crews report 1,200 barrels of oil recovered.

2000 Dispersant trial application conducted by C-130 with inconclusive results. Another test to be conducted on March 26 under optimal light conditions

2010 *Exxon Valdez* had transferred 11,000 barrels of oil to *Exxon Baton Rouge* before suspending oil transfer operations. Second test of ships cargo transfer system will be conducted on the morning of March 26.

2015 Exxon completes initial mapping of the area to determine the extent of the oil spread and impact. Bird rescue operation instituted by Exxon.

2045 Burn test conducted near Goose Island with approximately 100 square feet of tar left as a residue. An estimated 15,000 gallons of oil were consumed.

MARCH 26,1989 (THIRD DAY)

0530 CG Strike Team members join responders on *Exxon Valdez*. Earlier they had staged an Open Water Oil Containment and Recovery System (OWOCRS) for loading onboard USCGC *Sedge*.

0900 Exxon has assigned over 100 people including Exxon, Alyeska, and contract personnel to clean-up operation. Another 200 people are on standby.

0643 CG reports shipment of one MARCO skimmer from Elmendorf Air Force Base, Alaska. Twenty people transported to Bligh Island to survey beach cleanup and bird and wildlife impact. 75 oiled birds reported sighted by Department of the Interior (DOI) personnel. Birds include White Wing Scoters, Old Squaw, and Golden Eye ducks. Most oiled birds sighted on west side of Bligh Island. Two oiled sea otters sighted. DOI estimates that 3,000 seabirds and several hundred sea otters live in area of Knowles Head to Galena Bay.

1045 Three CG personnel arrive from MSO Anchorage to assist.

1115 CG personnel and members from ADEC and Exxon confer on state efforts for protecting sensitive areas on north side of Bligh Island.

1123-1510 USCGC *Sedge* arrives on scene north of Bligh Reef. USCGC *Sedge* dispatches small boat to recheck proper positioning of navigational equipment within the sound, maintaining navigation safety levels.

1400 Skimmers have recovered 2,275 barrels of oil from the water.

1500 National Transportation Safety Board (NTSB) representatives board *Exxon Valdez*.

1600 Dispersant application conducted with C-130, equipped with Beigart Air Deliverable Dispersant System (ADDS) system. Results are satisfactory.

1700 46,256 barrels of oil transferred to *Exxon Baton Rouge*.

1800 Skimmers recover 3,004 barrels of oil.

1830 Governor Cowper declares a state of emergency.

1900 51,064 barrels transferred to *Exxon Baron Rouge*. Commercial divers conduct survey with the use of video equipment.

2000 Additional survey by divers reveals port tanks intact. Some distortion is evident.

MARCH 27,1989 (FOURTH DAY)

High winds seriously impair response operations. Overnight, gusts clocked as high as 73 miles per hour have driven the spill nearly 40 miles into Prince William Sound, coating beaches at Little Smith, Naked, and Knight Islands. Skimmer systems, booms, and other equipment had to be moved to sheltered water for protection.

0542 USCGC *Rush* launches HH-65 helicopter to provide OSC and key response personnel an overflight view of the spill. USCGC *Rush* continues to enforce air space restrictions.

1100 90,599 barrels of oil have been transferred to *Exxon Baron Rouge.*

1245 CG overflight reveals that oil pooled up southwest of *Exxon Valdez* has formed dark, emulsified ribbons. Entire north and northeast side of Smith Island is heavily affected by thick oil. Some oil reaching Seal Island. Oil glancing eastern end of Naked Island is a lighter sheen, containing smaller ribbons. No recovery of oil since 1800, March 26.

1330 RRT assembles for a teleconference. The staging of sorbent materials at sensitive beaches is planned for areas in the path of the oil spill trajectory. Exxon is informed of the need for beach clean-up crews.

1900 124,299 barrels of oil have been transferred from *Exxon Valdez* to *Exxon Baton Rouge.*

CG overflight reports heavy shoreline contamination on northeastern end of Knight Island. Large pools of oil appear in the bays and inlets of the island. Oil heavily affects Eleanor Island moving west around north end of island. High winds move spill beyond sites initially selected for recovery operations

At the RRT teleconference, permission is given for dispersant application. Scheduled dispersant application did not take' place, because, aircraft did not arrive at authorized site. Virtually all mechanical recovery operations have ceased.

2100 3,000 barrels of oil recovered.

MARCH 28, 1989 (FIFTH DAY)

1000 Additional PACAREA Strike Team equipment arrives in Valdez.

1010 Exxon requests use of dispersants and in-situ burning around Eleanor Island (Zone 3). RRT considers dispersant use in this area inappropriate.

1030 USCGC *Sedge* informs OSC that no oil is in Main or Eshamay Bay. Work crews are setting booms in both areas.

1120 ADEC approves in-situ burning permit for area around Eleanor Island. PACAREA Strike Team reports 226,874 barrels transferred from *Exxon Valdez* to *Exxon Baton Rouge.*

1200 CG overflight from 0800 to 1130 indicates oil has reached beaches on all islands from northeastern side of Storey Island, Naked Island, Eleanor Island, Ingot Island, and Knight Island down to the Bay of Isles area on Knight Island. Heavy oil slicks are found between Naked and Eleanor Islands, extending in a westerly direction out into Knight Island passage.

Heavy oil impact on Smith and Little Smith Islands, with moderate contamination on the eastern side of Green Island. A major clean-up mobilization is initiated to protect critical fishery resources in Eshamay Bay, Main Bay, Port San Juan, and Esther Bay. Response management is reorganized formally as a steering committee consisting of CG, ADEC, and Exxon.

Major operation mobilized to protect fishery resources in Eshamay Bay, Main Bay, Port San Juan, and Esther Bay.

1230 Exxon-chartered C-130 applies dispersants in areas of heavily concentrated oil. Minor discharge of oil from *Exxon Valdez.* OSC authorizes use of dispersants.

1300-1530 Exxon dispersant operation conducted near vicinity of *Exxon Valdez.* Exxon reports excellent results. Dispersant operation also conducted off eastern end of South Island.

1400 OSC opens Port of Valdez to vessel traffic. Vessels subject to Vessel Traffic Service (VTS) regulations will make daylight transits only. A 1,000-yard safety zone around the *Exxon Valdez* is imposed. Tank vessels inbound or outbound will be required to have a two-tug escort to or from Bligh Reef. Vessels directed to avoid any clean-up operations by 500 yards.

1600 OSC holds teleconference with Alaska State Legislature.

1730 Secretary of Transportation Skinner, Environmental Protection Agency (EPA) Administrator Reilly, and CG Commandant Admiral Yost arrive to assess clean-up and oil transfer operations.

1800 274,000 barrels of oil transferred from the vessel at a transfer rate of 3,624 barrels per hour. Soundings indicate vessel is still hard aground between the number two and three starboard tank areas.

MARCH 29, 1989 (SIXTH DAY)

0845 Secretary of Transportation Skinner, EPA Administrator Reilly, CG Commandant Yost, Senator Murkowski, and Congressional staffers view spill area from CG overflight. They arrive in Valdez after having reviewed clean-up and oil transfer operations. They meet with OSC.

1200 At Valdez site, two dispersant aircraft stand ready for operations. An Aerostar aircraft augments command and control. Five landing craft (LCM) arrive for beach cleanup. Seven skimmers operate around Knight Island. Over 8,000 feet of additional boom is enroute. Skimmers remove over 5,000 barrels of oil.
Over 390,000 barrels of oil transferred off *Exxon Valdez.* Present pumping rate is 9,000 barrels per hour.

2000 Total of 442,988 barrels of oil-45 percent of cargo-now removed from *Exxon Valdez.* National Oceanic and Atmospheric Administration (NOAA) review of recent overflights indicates oil beginning counterclockwise rotation, moving from Naked Island, down western side of Knight Island, and up eastern side of Knight Island and western side of Green Island.

2130 *Exxon Baton Rouge* ceases taking on *Exxon Valdez* cargo. *Exxon San Francisco* is enroute to resume operations.

MARCH 30, 1989 (SEVENTH DAY)

Three separate beach clean-up work groups established. Priority Assessment Team to rank most critically affected areas for cleanup. Clean-up Assessment Team will determine use of best clean-up techniques. Shoreline Assessment Team will make final assessments of clean-up work.
Exxon reports that 7,537 barrels of oil recovered. NOAA estimates that 30 to 40 percent of the spilled oil has evaporated.
The oil has passed Montague Island and Latouche Island and is proceeding westerly into Gulf of Alaska.

0800 Oil concentrations have moved south seven to eight miles overnight, extending to the Montague Strait area near southern tip of Montague Island and eastern side of Latouche Island. Oil remains concentrated in center of Montague Strait. Green Island surrounded by oil. Heavy concentrations of oil remain in the area southeast of Naked Island, through the Eleanor and Ingot Island areas, and down western side of Knight Island. Vessel *Crystal Star* (130 feet) set up as Exxon floating command center.

0900 OSC authorizes three EPA members to assist Exxon in planning for beach cleanup.

0915 CG transportable communications center arrives from Sacramento, CA, to handle Coast Guard air traffic communications.

1000 Oil transfer from *Exxon Valdez* totals over 447,000 barrels.

1100 *Exxon San Francisco* takes over transfer operations from *Exxon Baton Rouge*. Divers in the water conduct additional surveys.

1300 VADM Robbins, Commander CG Pacarea, and Senator Lautenberg arrive in Valdez.

2300 *Exxon Baton Rouge* departs. Estimated 668,000 barrels of oil left on *Exxon Valdez* at conclusion of transfer operations.

MARCH 31, 1989 (EIGHTH DAY)

Due to low visibility, CG using its own AIREYE and Exxon infrared tracking equipment to monitor spill migration. NOAA projections using AIREYE surveillance shows large crossover effect. Oil slick appears to be turning back on itself, moving into Knight Island Passage. Oil emulsifying with water increases volume of liquid to be recovered. Responders replace weir-type skimmers with rope-mop units.

Green Island surrounded by thickening oil. Large patch of thick oil reaches northwestern side of Knight Island past Herring Bay area. Latouche Island touched by lighter patches. No signs yet of beach contamination. Exxon reports 7,537 barrels of oil recovered. Preliminary DOI survey of Green Island indicates 1,000 oiled birds.

1000 Senator Stevens arrives and is briefed by the OSC.

1143 USCGC *Sedge* in Sawmill Bay operating OWOCRS. Personnel are issued respiratory equipment to avoid irritating fumes from oil. Three dead sea otters removed from the sea.

1200 Nearly 80,000 feet of sorbent boom transferred from USCGC *Rush* to contract vessels in southern part of Prince William Sound. The *Rush* acts as command and control platform in this area. Operating OWOCRS from USCGC *Sedge,* Strike Force recovers 679 barrels of oil. . Meanwhile, *Exxon Valdez* shows signs of buoyancy after 500,000 barrels off-loaded to *Exxon Baron Rouge* and *Exxon San Francisco.*

1400 Alaska Air National Guard air drops sorbent materials to contractor boat crews in Hawkins Island area.

1700 Ten sea otters received by the wildlife cleaning facility.

1800 220,952 barrels of oil transferred from *Exxon Valdez* to *Exxon San Francisco.*

1930 USCGC *Sedge* departs for Sawmill Bay area by way of Knight Island passage.

APRIL 1, 1989 (NINTH DAY)

Heavy weathered oil continues to wrap around Knight Island. Emulsified oil reported from Squire Point south to Prince of Wales Passage opposite Port San Juan. Heavy oil also reported on west side of Latouche Island.

By April 1, a substantial accumulation of response equipment has been deployed throughout affected areas of the sound. For example, the amount of boom positioned by Exxon from March 24 to April 1 has grown from 12,500 feet to over 84,000 feet.

Galena Bay is protected by 1,000 feet of deflection boom; Head Main Bay by 5,000 feet, with a recovery vessel attending; Hatchery Island of Main Bay by 2,000 feet; and Herring Bay, Knight Island by 3,000 feet of sorbent boom and 6,000 feet of recovery boom used by five vessels for later pickup by skimmers.

Sawmill Bay, Evans Island protected by 50,000 feet of boom deployed with 15 vessels and much other equipment; Point Helens, Knight Island shielded by 1,200 feet of recovery boom; Snug Harbor, Knight Island by 1,000 feet; Bay of Isles, Knight Island by 500 feet of boom; and Bushby Island by 5,000 feet of recovery boom. Applegate Rock protected by CG skimming barriers with an attending 35,000-barrel recovery barge. Crippled *Exxon Valdez* is surrounded by 6,000 feet of boom.

The federal presence also has increased significantly. Employed in the response on April 1 are 391 CG personnel, 23 from DOI, 14 from NOAA, six from EPA, and four from the Department of Agriculture. On-scene equipment marshalled by federal agencies includes 8,000 feet of sea boom, 2,000 feet of flexi-boom, 1,200 feet of harbor boom, over 100,000 feet of sorbent boom, two CG Strike Force skimming OWOCRS, two Navy MARCO Class V skimmers, a PACAREA tow vehicle, eight boats, three CG cutters, four fixed wing aircraft, and four helicopters.

0650 295,645 barrels of oil transferred to *Exxon San Francisco.*

1000 State officials mobilizing resources to conduct water sampling in areas of hatchery and spawning activities.

1130 Wildlife recovery centers treating 28 oiled birds and 12 otters.

1215 Notice to mariners broadcast: all vessels not involved in response operations are to stay well clear of any observed oil.

1300 Secretary of Transportation Skinner and CG Commandant Admiral Yost briefed by OSC on cleanup status and adequacy.

APRIL 2, 1989 (TENTH DAY)

Using CG AIREYE, NOAA reports the leading edge of spill is approximately nine miles south of Cape Resurrection, progressing southwestward.

Beach cleanup at Naked, Peak, and Smith Islands begin as response teams gather growing clean-up manpower and pool special skills. Exxon team totals 160 persons now, including experts from the U.S., Canada, and the United Kingdom. The company has hired over 350 additional clean-up workers. Nearly 100 vessels are actively participating in the response.

336,853 barrels of oil have been transferred from the *Exxon Valdez* to *Exxon San Francisco.*

Exxon reports total quantity of oil recovered exceeds 10,000 barrels.

Exxon visual overflight indicates lighter sheens of oil are flowing into Gulf of Alaska. Large concentrations of oil remain in Knight Island Passage and in bays and sounds on north end of Knight Island.

ADEC beach surveys on Eleanor Island, Ingot Island, and northern end of Knight Island show heavy contamination.

0900 At wildlife cleaning centers, 28 oiled otters and 49 oiled birds are being treated. Approximately 140 oiled birds per square mile are found in Gibbons/Anchorage area. DOI estimates very high wildlife mortality rates.

1300 *Exxon San Francisco* loaded to its capacity of 452,533 barrels.

OSC grants Exxon request to apply dispersant to the slick sighted south of Point Erlington, but results not satisfactory on the main body of the oil. Dispersants moderately effective in breaking up surrounding oil sheen.

1400 Exxon reports 943,000 barrels of oil transferred from *Exxon Valdez* to *Exxon Baton Rouge* and *Exxon San Francisco*.

1530 *Exxon Baytown* alongside *Exxon Valdez* for continuation of oil transfer.

1600 150 birds treated and 30 sea otters recovered. Once treated, otters are transported to various aquariums.

By evening, southern extent of the spill progresses further south and is now 12 miles southwest of southern tip of Montague Island. Small stringers of oil sighted in the Bainbridge and Prince of Wales Passages. Light sheen remains in all passages. Three main streams of oil are flowing into Gulf of Alaska and are currently four to five miles offshore.

NOAA weather stations set up at northwest corner of Sawmill Bay and southern tip of Latouche Island.

APRIL 3, 1989 (ELEVENTH DAY)

Alaska Department of Fish and Game cancels all herring fishing in Prince William Sound based on damage to spawning areas.

1400 New remote weather stations established at northeast point of Sawmill Bay, Dangerous Island, and Perry Island.

1941 USCGC *Sedge* reports 8,949 barrels of oil recovered by CG OWOCRS.

2000 Skimming operations progress with oil recovery rates approaching 90 percent.

2100 Appearance of oil not yet sighted on shorelines west of Cape Puget, which serves as boundary between Valdez and Anchorage OSCs. Anchorage OSC sends representative to Prince William Sound. Preparations start to protect against possible movement of oil into Gulf of Alaska.

APRIL 4, 1989 (TWELFTH DAY)

Shoreline crews continue to operate at Smith and Naked Island. Housing for work crews provided by the barge *Exxon II*, located in Mummy Bay, Knight Island and by the M/V *Bartlett* (to be relieved later by M/V *Aurora*) in Sawmill Bay.

Exxon has established a boat cleaning station in Valdez. Cordova Fisheries Union are setting up another cleaning station in Cordova.

USCGC *Rush* works 329 aircraft and processed 1,867 radio contacts between 0700 and 2130.

Health and safety training classes set up by Exxon contractors to provide mandatory training for all clean-up personnel.

Oil transfer from *Exxon Valdez* completed by *Exxon Baytown; Exxon Baytown* underway.

APRIL 5, 1989 (THIRTEENTH DAY)

Over 66,000 feet of boom deployed in Sawmill Bay. This represents 65 percent of total boom deployed. OSC had decided to deploy a significant amount of booming and skimmers in defensive positions to protect hatcheries, removing capacity to fight the spill itself.

Air Force Military Airlift Command (MAC) airlifts U.S. Navy, CG, and Exxon skimmer boats, drums of dispersant, mooring systems, boom vans, barrier material, and assorted vehicles from California, Oregon, Texas, Virginia, Denmark, and Finland to spill site. Exxon will pay airlift cost.

Sawmill Bay fishermen now expressing confidence that hatcheries will be protected. Clean-up crews at a number of beach locations begin to mop up oil in tidal pools.

Primary concentration of oil in Prince William Sound extends almost in a continuous sheen from Smith Island, between Knight and Green Islands, and down Montague Strait out into Gulf of Alaska. Oil remains in passages between Bainbridge and Latouche Passages. Large slick has moved into Gulf of Alaska, extending from southern outlets of the passages across Montague Strait. Slick has moved approximately 50 miles into the gulf.

Skimming rates continually reduced due to oil weathering.

OSC permits transit of two vessels during daylight hours, provided they are heading in the same direction. Each vessel must have two-tug escort.

USCGC *Rush* works 127 air contacts and processes 550 radio contacts as part of air traffic control operations.

1035 *Exxon Valdez* refloated after oil transfer operations and is holding position on Bligh Reef. About 16,445 barrels of oil remain in the vessel. 1,000-yard safety zone established around the tanker. Two MARCO skimmers and a vacuum truck are aboard *Exxon Valdez* and attended by workboats and standby dispersant-loaded aircraft to respond to any additional spill. Transit of *Exxon Valdez* to Naked Island area begins.

1420 USCGC *Sedge* reports recovery of oil is becoming extremely difficult due to formation of a water-and-oil emulsion or 'mousse.'

1935 *Exxon Valdez* anchored in Outside Bay near Naked Island.

2201 Canadian and U.S. Region Joint Contingency Plan activated. Co-chairs are Captain G.E. Haines, the Commander of the Coast Guard District 17 Marine Safety Program, and Mr. G.R. Stewart, Director General, Western Region, Canadian Coast Guard.

APRIL 6, 1989 (FOURTEENTH DAY)

Twenty-one additional skimmers, including 15 Navy units transported by the Department of Defense (DOD), enroute to spill scene.

Variety of response equipment being assembled at Mummy Bay and Point Helens, Knight Island to protect environmentally sensitive areas there. Also, joint U.S. - Canadian response plan invoked to speed delivery of more clean-up equipment and operators.

USCGCs *Midgett, Yocoa, Sweetbriar, Iris,* and *Planetree* directed to join cutters *Rush, Ironwood,* and *Sedge* in clean-up area.

Mandatory health and safety classes for all clean-up crews begin while contractors work with NOAA to develop detailed maps of oiled beach areas.

Exxon Valdez remains anchored off Naked Island.

NOAA overflight reveals oil is thinning and heading out to sea. Heavy oil contamination reported at Smith Island, Main Bay, Falls Bay, Eshamay Bay, eastern side of Chenega Island, and northern parts of Bainbridge, Evans, and Latouche Islands. Eastern and western shores of Knight Island also contaminated. Oil with light concentrations of emulsified ribbons spotted north of Main Bay near Port Nellie Juan.

Mortality rate of otters turned into rehabilitation centers is approximately 50 percent. Leading edge of oil slick 22 miles south of Nuka Bay in Gulf of Alaska. Impact observed on the Chiswell Islands. Oil mousse surrounds Barwell Island, and some oil has been trapped on eastern side of Cape Resurrection.

Oil observed approximately 20 miles off coast from Gore Point, varying in width from 10 to 20 miles. Oil forming wind rows. Oil in Prince William Sound continues to flush into the gulf.

Rear Admiral Nelson (USCG) assumes OSC responsibility to facilitate strategic control of response.

Oil volume in Bainbridge and Latouche Passages diminishes with migration of spill into the gulf.

USCGC *Rush* maintaining air traffic control.

Oil affects areas within Anchorage OSC jurisdiction. Oil slick reaches Barwell Island at entrance of Resurrection Bay.

1500 Overflights indicate difficulties encountered in positioning skimmers in areas of heavily concentrated oil. Emulsified patches of oil clinging to some shoreline areas inaccessible to larger skimmers.

APRIL 7, 1989 (FIFTEENTH DAY)

At direction of President Bush, DOD establishes Director of Military Support Joint Task Force (DOMS JTF) to assist OSC in cleanup. DOD assessment team will determine best way to apply military support. Joint Task Force begins daily oversight meetings in Pentagon Army Operations Center.

Emergency order tightening operations at Valdez Terminal signed by Governor Cowper.

Spill area enlarged to approximately 2,600 square nautical miles, according to NOAA analysis of recent overflights. Heavy concentrations of oil sighted on eastern side of Knight Island. Sheen remaining in most passage areas forms streams and stringers.

Approximately 300 dead birds and 76 sea otters collected. The new Valdez rehabilitation center begins operations.

0930 Divers survey tank number 1C beneath *Exxon Valdez* and begin drilling operations to prevent further spread of main crack.

1830 Sheen with streaks of mousse reported extending from the northern part of Naked Island down eastern shore of Knight Island, through Latouche Passage, and into Gulf of Alaska. Northern part of Montague Island and Green Island affected. Extensive sheen observed in and around Snug Harbor. Light sheen with stringers ranges from Port Nellie Juan to Main Bay and down Knight Island Passage into Latouche Passage. Some sheen observed in Prince of Wales Passage.

USCGC *Rush* maintains air traffic control. USCGC *Sedge* passes OWOCRS towline to *Theresa Marie.*

APRIL 8, 1989 (SIXTEENTH DAY)

Skimming operations continue in Main, Eshamay, Herring, and Sawmill Bays and begin between Knight and Green Islands. While GT-185 skimmers are very effective, CG reports that oil recovery rates are reduced to 200 barrels per day due to increasing oil viscosity.

Morning overflight shows occasional light sheen in Perry Passage north of Port Nellie Juan. No oil found in Wells Passage or McClure Island. Light sheen at Port Nellie Juan, Main Bay, and Crofton Island where some beach impact observed. Less than 10 percent of Eshamay Bay covered with sheen, but heavy oil concentrations contained by booms.

Mixture of sheen and mousse observed in northern, eastern, and southern areas of Knight Passage. Sheen and mousse streaks noted in Prince of Wales Passage. Shorelines abutting Latouche Passage show oiling with mousse and sheen offshore. Sheen with patches of heavy oil observed off southern Montague Island. Trajectory of spill curving into Anchorage OSC jurisdiction. Valdez and Anchorage OSCs deploy MARCO Class V skimmers in defensive positions in Gulf of Alaska.

DOI reports that 529 birds and 94 sea otters have died. The Fish and Wildlife Service (FWS) has prepared a list of wildlife areas believed to be at risk from the oil spill.

USCGC *Rush* maintains air traffic control, works 303 aircraft, and processes over 2,126 radio contacts. USCGC *Ironwood* continues installation of mooring systems in Sawmill Bay.

CG Boating Safety Team enroute to Whittier to conduct safety boardings on volunteer recreational boats used for retrieval of dead wildlife.

In Whittier, DOI is setting up a wildlife collection station and Exxon establishes a boat-cleaning station.

0630 Overflights conducted by NOAA.

0945 USCGC *Rush* reports heavy concentration of oil from Bass Harbor to eastern end of Smith Island. Slick apparently one mile wide. Exxon officials notified and a skimmer crew diverted.

Response actions stepped up at Sawmill Bay and Snug Harbor, where eight skimmers, five vessels, and an oil recovery barge are involved. Over 5,000 feet of boom deployed in King Bay. Fully-boomed *Exxon Valdez* remains anchored off Naked Island with a 32-foot water cushion for each of its damaged tanks. Light sheen reported inside the boom.

FWS conducts aerial shoreline survey from Prince William Sound to Homer.

2130 Captain Ryan of the Canadian Coast Guard says Canadian skimming equipment has recovered 666 barrels.

2200 USCGC *Sedge* enroute to Snug Harbor.

2205 Summary of clean-up activity provided by Exxon as of this date:

Initial Amount of Oil Spilled:	240,000 barrels
Amount Recovered:	17,000 barrels
Amount Evaporated:	77,000 barrels
Amount Dispersed:	11,000 barrels
Amount in Prince William Sound:	45,000 barrels
Amount in Gulf of Alaska:	45,000 barrels
Amount on Beaches:	45,000 barrels

CG AIREYE overflights reveal scattered mousse and sheen from Cape Junken to the southern section of Otter Island. Some oil apparent around Chiswell Island. Light sheen with thin strands of mousse apparent north of Hinchinbrook Island. Light beach impacts were observed on the northeastern part of Montague Island.

Exxon divers complete drilling eight stopper holes in *Exxon Valdez* to arrest fractures.

APRIL 9, 1989 (SEVENTEENTH DAY)

Spill seems to be stabilizing. CG reports leading edge of the spill has not advanced in two days. Sheen with streaks of congealed oil extends from northern Naked Island through the lower passages and into Gulf of Alaska. Heavy sheen reported around Snug Harbor.

Sawmill and Herring Bays and Snug Harbor continue to hold highest response priority. Arriving at Sawmill Bay to join the response effort are a floating hotel (housing 1,000 response personnel), five waste-oil barges, five waste-oil "doughnuts," and 100 small skiffs.

USCGC *Rush* works 320 aircraft and has processed 2,180 radio contacts in the last 24 hours.

1430 VADM Robbins, CG PACAREA Commander, returned to Valdez.

1515 Joint NOAA/USCG overflight reports no oil found on Hitchinbrook Island near the Hawkin's Island cut-off.

1945 Aerial reconnaissance reports leading edge of spill is 25 miles southeast of Nuka Island. Slick runs close to the shore from Cape Junken to the vicinity of Resurrection Bay, where fresh water runoff and fjord winds are pushing the spill offshore.   Major spill impact observed in Chiswell Islands due to combination of steep shoreline and high wave energy. Offshore slick appears as 20 to 30 mile sheen with widely separated areas of mousse.

Weather hinders clean-up operations. Many skimmers operating in exposed areas head for more protected waters.

CG air operations total 38 hours of flight time.

In the OSC Anchorage-Gulf of Alaska operational theater, the USCGC *Yocona* has sailed from Kodiak to Seward and presently is in Seward. A Navy MARCO skimmer is on scene. An 84-inch boom at Seward cannot be deployed by USCGC *Yocona* and *Planetree* due to adverse weather conditions offshore.   Test using herring nets to break up areas of oil considered partially successful.

Two 65,000-gallon capacity bladders enroute to Seward for use in skimmer operations.

USCGC *Morgenthau* is stationed at the southeast entrance to lower Cook Inlet monitoring traffic.   Remote weather stations are planned for Barwell Island, Outer Island, Chugach Island, and Marmot Island.

14,000 feet of boom deployed at Resurrection Bay and Kenai Fjords National Park.

APRIL 10, 1989 (EIGHTEENTH DAY)

Leading edge of the slick located 20 miles south of Nuka Sound.

USCGC *Sedge* conducts shoreline survey of Snug Harbor with a small boat. Only a very light sheen observed approximately one mile offshore. Four foot wide band of black oil observed at the high water mark on the beach.

FWS personnel continue shoreline aerial survey from Prince William Sound to Kodiak. On the ground, a survey of deceased wildlife is conducted on the north end of Knight Island.

USCGC *Rush* works 324 aircraft, processes 2,192 radio contacts, and then puts into port for logistic resupply. USCGC   *Rush* is relieved of air traffic control responsibilities by USCGC   *Sedge*.

Poor visibility and high variable winds hamper overflight assessments.

1400 USCGC *Ironwood* enroute from Snug Harbor to Valdez and reports every five miles on concentrations of oil.   *Ironwood* reports light sheen 500 to 1,000 yards wide between Sleepy Bay and Point Helen. Several ribbons of oil approximately 10 by 420 yards reported three miles south of Discovery Point.

1845 USCGC *Storis* loaded with approximately 2,000 feet of membrane-type boom for transport to Kitoi fish hatchery at Afognak Island.

At OSC Anchorage-Gulf of Alaska zone, gale force winds and 20-foot seas prevent offshore operations. Exxon establishing an otter cleaning station in Seward. 2,100 feet of boom are deployed at Tutka Bay fish hatchery.

Two Navy MARCO skimmers with CG bladder arrive at Homer.

Six fishing vessels equipped with herring nets depart Kodiak to join 30 other fishing boats at Seward to form mobile response unit in attempt to break up oil patches off Cape Resurrection. Additional 10,000 feet of boom will be deployed with these vessels.

APRIL 11, 1989 (NINETEENTH DAY)

Total of nearly 200,000 feet of boom have been deployed to protect endangered areas in Prince William Sound. Included are 85,000 feet of containment boom, 98,000 feet of absorbent boom, and 12,000 feet of boom surrounding *Exxon Valdez* anchored off Naked Island.

39 skimming operations shut down by rough seas on April 10 have yet to resume operations in the natural collection area of Snug Harbor, Sawmill Bay, Point Helen, Latouche Pass, and Herring Bay. Over 80 people are now involved in the cleanup of Naked Island, with 500 more workers expected to join clean-up efforts by April 13.

Oil moved westward, forming long, well-defined bands of mousse along eastern shore of Latouche, Knight, and Ingot Islands. Light sheens observed in Main and Eshamay Bays.   Large patch of sheen/mousse combination approximately six miles long and two miles wide observed west of Eleanor Island, extending nearly into entrance of Main Bay.

Small amounts of sheen and mousse observed in Perry Island area. Herring Bay still heavily oiled. Large band of sheen and mousse parallels southeast end of Knight Island. Bay of Isles beaches also heavily oiled. Wind and wave conditions over past two days have mixed and dispersed the larger concentrations of oil in open waters of Prince William Sound.

Exxon divides Prince William Sound into four quadrants, each with a command and control vessel (with PACAREA Strike Team member aboard) to coordinate oil recovery operations. Quadrant zone one represents area north of Knight Island, Quadrant zone two represents areas west, and Quadrant zone three areas east. Quadrant zone four represents areas south of island. Areas other than Prince William Sound are designated zone five.

Exxon Valdez is subject of diving survey. Repairs made to fractures in hull of number four starboard tank.   Vessel engines are checked   **and** considered operational.   To date, 19,000 barrels of oil recovered, but bad weather hinders future recovery operations.

0340 USCGC *Morgenthau* ordered to mouth of Resurrection Bay to coordinate efforts to break up oil by fishing vessel fleet and Exxon spotter plane.

1445 Potato Point Radar Site becomes inoperative, forcing OSC to close the Valdez Narrows to vessel traffic.

1500 Radar site reactivated and port reopened.

APRIL 12, 1989 (TWENTIETH DAY)

CG helicopter overflight reports leading edge of slick located 30 to 40 miles southeast of Gore Point.

Snug Harbor and the Bay of Isles relatively free of oil, but shorelines are extensively oiled.

1535 ADM Yost, Commandant USCG, arrives at Elmendorf AFB, Anchorage. He returns as the President's and the Secretary of Transportation's representative to oversee the spill cleanup. He is met and briefed by VADM Robbins and RADM Nelson.

APRIL 13, 1989 (TWENTY-FIRST DAY)

CG monitor and vessels assigned to each of five designated clean-up sectors. Dispersant tests show no effect on mousse and little effect on sheen.

1122 ADM Yost meets with Exxon officials to establish clean-up priorities. Exxon tasked with submitting a beach clean-up work plan. Exxon temporarily suspends shoreline cleanup pending submission and approval of the plan.

1330 RRT meeting held.

1800 Transfer of oil slops from T/S   *Exxon Valdez*   completed. Internal survey underway.

1900 ADM Yost briefs operations committee on his purpose and function as President's representative.

APRIL 14, 1989 (TWENTY-SECOND DAY)

1115 Commandant meets with Governor Cowper and Commissioner of ADEC to discuss clean-up progress and strategies. Commandant also briefed by LT GEN McInerney on results of DOD assessment team study.

1300 RRT meeting held.

1314 ADM Yost provided status report during telecom with President Bush. The effectiveness of hot water/steam cleaning of shoreline discussed, and a status report provided by ADM Yost.

1500 LT GEN McInerney, ADM Yost, and VADM Robbins meet. DOD support resources discussed and additional resources secured, including COE Dredge *Essayons* and U.S. Navy Ship *Juneau*.

1700 ADM Yost meets with top Exxon officials. He presents them with a list of 50 beaches requiring cleanup. ADM Yost is putting pressure on Exxon to provide additional personnel within 10 days.

2000 ADM Yost meets with SSC and operations committee. The need to protect Seward, Homer, and Kodiak is discussed, as is forthcoming Exxon shoreline clean-up plan.

2100 Shoreline Clean-up Committee approves use of wash-vacuum oil cleaning system (VIKOVAK) on eastern shore of Smith Island. Committee also approves test cleaning using hot/cold/high-pressure water flushing with VIKOVAK applications on northern portion of Smith Island. Instructions given to avoid all living spicies, backshore and upper intertidal areas, and use of high-pressure water or steam where invertebrates and seaweed exist.

APRIL 15, 1989 (TWENTY-THIRD DAY)

0905 ADM Yost, VADM Robbins, and NOAA rep conduct overflight of Northwest Passage to observe skimming.
Concentrated skimming operations continue in many areas. Ten skimmers and eight CG cutters operating on scene.
RADM Nelson departs to resume duties as Commander of Seventeenth CG District.

1400 Hydrovac pumping system transferred to Zone two to speed offloading of skimmers there. Hydrovac systems considered only effective pumping system for the viscous, debris-laden oil that is difficult, to transfer through integrated skimmer pumping systems.

1600 Vice Admiral Robbins (USCG) becomes federal OSC (FOSC).

1700 Commandant meets with representatives of Exxon, ADEC, and fishermen. Exxon presents shore clean-up execution plan.

1900 FOSC approves shoreline clean-up work order for Eleanor Island.

1900 ADM Yost attends operations committee briefing to discuss the day's developments and clean-up actions to be taken outside Prince William Sound.

APRIL 16, 1989 (TWENTY-FOURTH DAY)

0930 Commandant, FOSC, and several federal, state, corporate, and press representatives attend shoreline washing experiment and demonstration conducted by Exxon on southwest Eleanor Island.

Exxon submits shoreline clean-up execution plan to FOSC and staff for review.

Prince William Sound overflight shows significant change in the path of oil caused by changes in wind direction. Large concentrations of mousse and sheen previously seen near Eleanor and Ingot Islands now being driven southwest towards Falls and Main Bay and Lone, Perry, and Culross Islands. Significant shoreline impact anticipated there. Projections indicate oil will remain in that vicinity and will not migrate into Wells Passage or Port Nellie Juan. Overflight conducted in the area of Gore Point shows shoreline impacts. Remaining oil in the gulf between Cape Junken and the Chugach Islands may be driven northward and may reach shoreline in that area due to predominantly southeast winds.

Clean-up operations temporarily stopped due to reports of exposures to harmful vapors. Air quality monitoring shows exposure limits within safety guidelines.

1800 ADM Yost meets with operations committee. Alternate methods of beach cleanup demonstrated earlier discussed and evaluated. ADM Yost expresses concern over the high number of skimmers that he observes not operating during his earlier overflight.

APRIL 17, 1989 (TWENTY-FIFTH DAY)

Skimmer operations are redirected in order to concentrate on near-shore areas to recover larger amounts of accumulated oil more effectively.

Salvage of *Exxon Valdez* continues. Box patches installed from frame 1 to fore and aft bulkhead. All 'hangers: (hanging steel pieces) removed from tank numbers 1C, 2C, and 3C. Divers conduct survey of tank number 1S.

Joint command communications network established to connect primary command nodes for overall operations coordination.

1000 Commandant meets with lead agencies to discuss comments on Exxon workplan.

1300 Commandant approves workplan and presents comments to Exxon officials.

1400 Captain Crowe (USCG) assumes duties as Chief of Operations.

1600 Captain Roussel (USCG) designated assistant OSC for spill outside Prince William Sound. Vice Admiral Robbins remains FOSC for entire spill.

1900 ADM Yost and executive committee meet. He urges immediate pursuit of shoreline cleanup using acceptable methods.

APRIL 18, 1989 (TWENTY-SIXTH DAY)

1315 ADM Yost, FOSC, and RADM Baker, USN, Commander Third Amphibious Group discuss naval support of cleanup.

1330 Visiting Florida DNR personnel discuss cleanup with FOSC.

1400 ADM Yost departs Valdez for Anchorage.

Skimming operations center on heavy concentrations of oil near Perry, Long, and Culross Islands.

Nine CG Cutters operating in area. Total of 53 vessels, including 33 skimmers on scene.

Prince William Sound overflight reveals heavy patches of oil from Wells Passage to Lone Island. Oil sheen collecting in Port Nellie Juan. Light winds keeping oil basically immobile. New light oil sightings on south side of Lone and Eleanor Islands and south to Smith Island. Overflights continue to show sheen and mousse patches from Chugach Islands east to Cape Resurrection. Sheen and mousse sighted in vicinity of Shuyak Island. No oil sighted along beaches of Katmai National Monument.   Very light tar ball splattering Seward's 2,000-foot beach front. No other evidence detected.

Soviet skimmer M/V *Vaydagursky* receives approval for thirty-day entry into U.S. waters. Approval includes authorization to work within three miles of land from Valdez to Homer and around the Kodiak Archipelago, and to make port calls.

APRIL 19, 1989 (TWENTY-SEVENTH DAY)

0700   Two Navy MARCO Class V skimmers, two Class XI skimmers, and two Exxon contract skimming vessels deployed at leading and trailing edge of heavy oil concentration in Perry Passage.   Five Navy MARCO Class V skimmers and Exxon contract skimmers are deployed in the bays west of Eleanor Island to collect oil pushed by westerly winds. Three Navy MARCO Class V skimmers deployed west of Squire Island.

1000 USCGC *Sweetbrier* on scene at Esther Island hatchery to deploy SUPSALV mooring system for protective booming operations.

1100 ADM Yost and FOSC attends luncheon with mayors of affected towns prior to ADM Yost's departure from Anchorage.

1200 Soviet M/V *Vaydagursky* skimmer vessel arrives at Seward. CG representatives board vessel with interpreter, pilot, and VECO representatives (Exxon contractor) to discuss proposed operations. Vessel currently refueling and preparing for skimming operations near mouth of Resurrection Bay.

Salvage operations continue.   All tanks except number 4S are inerted. Tank number 4s is opened and safe for work. Three box patches installed in the tank over the small 'repaired' fractures. Tank number 4s has been repaired temporarily.

1300 FOSC meets in Homer with RRT to discuss shoreline cleanup.

APRIL 24, 1989 (TWENTY-EIGHTH DAY)

0930 FOSC briefs Secretary of Interior Lujan, and Congressmen Young, Galley and Weldon.

Shoreline activities continue on Naked Island. Over 250 Exxon contractor personnel are supporting clean-up operations in four areas on north side of Naked and Peak Islands.

Nearly 41,062 barrels of oil and water mixture and 14,270 barrels of oil transferred to a barge alongside *Exxon Valdez*.

State of Alaska-funded and constructed Joint Communication Center added to the response communications network.

Army Corps of Engineers (ACOE) dredge *Yaquina* deployed as a skimmer, recovers 1,100 barrels of oil.   Total of 53 vessels, including 35 skimmers, are performing oil recovery operations.

APRIL 21, 1989 (TWENTY-NINTH DAY)

1300 RRT teleconference updates members and tasks them w:th investigating use of COREXIT 7664 dispersant for shoreline cleanup.

Air Force has transported total of 928 tons of response equipment by 15 C-5 missions and 97 tons of equipment by four C-141 missions. Salvage operations to cut hangars from bottom of *Exxon Valdez* continue. Draft proposals for tank and hull cleaning presented to CG.

High volume/low pressure beach washing applied by 250 Exxon personnel to Naked Island shoreline. CG reports 240 feet of shoreline cleaned.

Skimmers move into Knight Island passage area for oil recovery. MARCO skimmer working in Sawmill Bay in concert with CG monitor. OWOCRS and 84-inch boom deployed in Resurrection Bay.

M/V *Vaydagursky* with CG monitor onboard working with two tugs to rig skimming booms in Resurrection Bay. 58 vessels, including 37 skimmers, are operating as recovery or support craft.

APRIL 22, 1989 (THIRTIETH DAY)

Divers continue to cut hanging steel pieces from *Exxon Valdez* hull. Stopper holes are drilled at ends of all transverse fractures. ADEC personnel scheduled to arrive on M/V *Winter King* to monitor water quality and observe repairs. Some oil continues to surface from bilge keels where it is trapped.

Shoreline clean-up plan for Applegate Rock area approved by FOSC. Land use permit for clean-up operations on all state-owned tide and submerged lands received from State of Alaska for 1989.

Skimmers in Resurrection Bay unable to pump debris-contaminated, weathered oil. Mr. Clean Class III skimmer recovers oil/mousse off Gore Point. USSR skimmer *Vaydagursky* shut down for modifications. It had recovered 12 barrels of oil.

APRIL 23, 1989 (THIRTY-FIRST DAY)

Skimming operations continue to maximum degree possible in Prince William Sound Upper Passage, Northwest Bay, Lower Knight Island Passage, and Eshamay Bay. Exxon estimates 2,990 total barrels recovered on April 22. Adverse weather forces halt of skimming operations.

Vessel with 4,800 feet of boom and U. S. Navy MARCO skimmer dispatched to Kitoi Hatchery, Izhut Bay. Eight fishing vessels from Seldowa and Port Graham deployed to Flat Island. They are towing herring nets in attempt to collect mousse and tar balls in area.

Bird cleaning station opens at National Guard Armory and otter station at National Marine Fisheries Services site, Gibson Cove. Alaska Department of Fish and Game closes herring fishery on north and west side of Afognak Island due to sheens in area. Boat cleaning station operational at Herman's Harbor,

CG and Exxon representatives visit villages on Kodiak Island to gather and disseminate information.

APRIL 24, 1989 (THIRTY-SECOND DAY)

Cutting hangers from *Exxon Valdez* completed. Drilling stopper holes at end of fractures continues. Stripping Forepeak and number 1C tank to be completed. Oil in number 1C tank to be boomed to reduce oil leaks. ADEC personnel on M/V *Winter King* alongside *Exxon Valdez*.

Adverse weather continues to hamper efforts to skim oil. ACOE Dredge *Yaquina* with CG skimming barrier manages to operate in South Night Island Passage and Mummy Bay. Two SUPSALV skimmers work in Northwest Bay. Remainder of 58 vessels, including 37 skimmers, stay in sheltered waters. Two hundred feet of shoreline cleaned during last two days.

CG and Exxon personnel brief Senator Stevens.

APRIL 26, 1989 (THIRTY-THIRD DAY)

Drilling of stopper holes continues on *Exxon Valdez.* Booming of oil in tank number 1C in progress.

Adverse weather continues to hamper skimming. All western Alaska skimmers, except for Mister Clean III, attempting to get to Division Bay, Naka Passage to skim oil/mousse concentrated in that area. Approximately 15,000 feet of boom arrive in Homer and will be distributed as needed in area by MAC Group. A total of 58 vessels continue to be involved in operation, but 42 of these now are skimmers.

Joint Communications Center now operational. Phone patch capability of this system allows total interconnections among all deployed units ashore or afloat.

1900 FOSC attends evening operations meeting with Exxon personnel onboard the U.S. Naval Vessel *Juneau.*

APRIL 26, 1989 (THIRTY-FOURTH DAY)

Tank cleaning and repair activities on *Exxon Valdez* continue. Four major networks give coverage to existence of Coast Guard tapes of radio conversations with *Exxon Valdez* at time of grounding.

Clean-up operations in western Gulf of Alaska continue to be hindered by adverse weather. Due to debris-laden, weathered condition of recovered oil, offloading in both western Gulf of Alaska and Prince William Sound is slow and difficult. Various super-suction devices have been tried with limited success. The operation has been enhanced by heating the oil with stem coils, but it takes two to two-and-a-half hours to heat approximately 32 barrels.

300 feet of shoreline cleaned in Northwest Bay by April 25. Multiagency monitoring program is established to ensure that all shoreline segments will **be** cleaned in the presence of a federal and state monitor.

58 vessels remain in clean-up operation; 42 are skimmers.

1730 CAPT Calhoun, USCG, CO, MSO Portland, OR. arrive on scene to survey *Exxon Valdez* damage and condition.

APPENDIX B. RESPONSE FORCES ON SCENE

Exxon and U.S. Coast Guard Response Forces On Scene

MARCH 24	MARCH 25	MARCH 28	APRIL 02	APRIL 07	APRIL 12
Exxon	**Exxon**	**Exxon**	**Exxon**	**Exxon**	**Exxon**
10 Landing Craft (mechanized) 15 Vessels (various) 12,500 ft of Boom 3 Skimmers 164 Personnel	52 Vessels (various) 26,000 ft of Boom 6 Skimmers 237 Personnel	71 Vessels (various) 34,000 ft of Boom 7 Skimmers 340 Personnel	107 Vessels (various) 110,000 ft of Boom 12 Skimmers 817 Personnel	110 Vessels (various) 158,000 ft of Boom 25 Skimmers 18 Aircraft 1046 Personnel	210 Vessels (various) 283,000 ft of Boom 41 Skimmers 25 Aircraft 1300 Personnel
Alyeska Marine Terminal	**U.S. Coast Guard**	**U.S. Coast Guard**	**U.S. Coast Guard**	**U.S. Coast Guard**	**U.S. Coast Guard**
4 Boats 1 Class V Skimmer 1 Class VII Skimmer 1 Sea Skimmer 15,200 ft of Boom 1 Helicopter 3 Sea Pac Barrier Systems 28 Personnel **U.S. Coast Guard** PACAREA Strike Team (11 people)	1 High Endurance Cutter 1 Helicopter 1 Buoy Tender 1 32-ft Boat PACAREA Strike Team (13 people) 2 Strike Team Air Deployable Anti-Pollution Transfer Systems (ADAPTS) 38 Personnel at Marine Safety Office Valdez	1 High Endurance Cutter 1 Helicopter 1 Buoy Tender 1 32-ft Boat 1 USCG AIREYE FALCON Plane 1 USCG C-130 Transport Plane PACAREA Strike Team (16 people) 5 ADAPTS 1 Strike Team Open Water Oil Containment and Recovery System (OWOCRS)	1 High Endurance Cutter 4 Helicopters 2 Buoy Tenders 1 32-ft Boat 1 USCG AIREYE FALCON Plane PACAREA Strike Team (20 people) 6 ADAPTS 2 OWOCRS 11,200 ft of Boom 96 Personnel at Marine Safety Office, Valdez and Anchorage	1 High Endurance Cutter 6 Helicopters 2 Buoy Tenders 6 Small Boats 3 Fixed Wing Planes PACAREA Strike Team (20 people) 6 ADAPTS 2 OWOCRS 11,200 ft of Boom 118 Personnel at Marine Safety Office Valdez and Anchorage	2 High Endurance Cutters 2 Medium Endurance Cutters 4 Buoy Tenders 6 Small Boats 3 Fixed Wing 6 Helicopters PACAREA Strike Team (20 people) 6 ADAPTS 6 OWOCRS 15,200 ft of Boom 208 Personnel at Marine Safety Office Valdez and Anchorage

Other Federal Response Forces On Scene

MARCH 24	MARCH 25	MARCH 28	APRIL 02	APRIL 07	APRIL 12
	U.S. Navy		**U.S. Navy**	**U.S. Navy**	**U.S. Navy**
	2 Skimmers 4 Tow Boats 2 Support Vans		5 Skimmers 6 Tow Boats 8 Support Vans 14 Mooring Systems	20 Skimmers 2 Barrier Skimming Systems 10 Tow Boats 20 Mooring Systems 11 Support Vans 94 Personnel	20 Skimmers 2 Barrier Skimming Systems 10 Tow Boats 20 Mooring Systems 11 Support Vans 94 Personnel
		U.S. Forest Service	**U.S. Forest Service**	**U.S. Forest Service**	**U.S. Forest Service**
		2 Personnel	6 Personnel	23 Personnel	30 Personnel 1 Helicopter
NOAA	**NOAA**	**NOAA**	**NOAA**	**NOAA**	**NOAA**
1 Helicopter 6 Personnel	1 Helicopter 10 Personnel	1 Helicopter 13 Personnel	1 Helicopter 14 Personnel	1 Helicopter 12 Personnel	1 Helicopter 4 NOAA Data Buoys 22 Personnel
		FAA	**FAA**	**FAA**	**FAA**
		7 Personnel	7 Personnel	7 Personnel	7 Personnel
Dept. of the Interior	**Dept. of the Interior**	**Dept. of the Interior**	**Dept. of the Interior**	**Dept. of the Interior**	**Dept. of the Interior**
2 Personnel	2 Personnel	21 Personnel	23 Personnel	23 Personnel	26 Personnel
	EPA	**EPA**	**EPA**	**EPA**	**EPA**
	2 Personnel	2 Personnel	6 Personnel	6 Personnel	7 Personnel
					National Guard
					93 Personnel

Source: U.S. Coast Guard, 1989.

ADAPTS Air Deployable Anti-Pollution Transfer System
ADAPTS is a pumping system that is transportable in a variety of ways. It is designed for rapid deployment to dewater, remove oils, or transfer a limited number of liquid hazardous materials. A Type III Avco Lycoming diesel prime mover powers the ADAPTS to push oil through a discharge hose for transfer into a suitable container. ADAPTS is operated and owned by the Coast Guard.

AIREYE AIREYE is an aerial surveillance, information gathering, and recording system installed on certain Coast Guard aircraft. This state-of-the-art system employs multiple information inputs (visual-photographic, infrared radar) to monitor and track surface objects. AIREYE usually is operated on marine safety and environmental monitoring missions. It has law enforcement and military applications.

BOOMS Booms are primarily barriers used to contain or deflect an oil slick, or prevent oil from reaching an environmentally sensitive area.
There are two general types of booms: fence booms and curtain booms. Fence booms are constructed from rigid or semi-rigid material and serve as a vertical barrier against oil floating on water. Curtain booms have a flexible skirt that is held down by ballasting weights or a separate tension line. Booms often are referred to in terms of their combined length below (draft) and above (freeboard) the waterline. Therefore, an 18-inch boom would be one that has a total barrier height above and below the waterline of 18 inches.
Booms are designed for special purposes. Fire booms consist of fireproof material to contain oil for in-situ burning. Sorbent booms absorb oil on contact and are disposable. Ice or cold weather booms can withstand extreme temperatures and ice flows. Some booms have skimming capabilities that trap oil for recovery.

GT-185 The GT-185 is a weir skimmer utilizing gravity as a means of collecting oil. With its edge placed at the waterline, the body of the skimmer below water forces oil into a pump. Oil then is conveyed to a suitable container. The GT-185's pumping system can deliver up to 440 gallons per minute.

MARCO SKIMMERS The MARCO skimmer is a sorbent lifting-belt skimmer that works on a conveyor belt-type system. The lifting belt has a high affinity for oil that prevents large volumes of water from being recovered along with the oil. Oil is scraped or wrung from the belt into an appropriate container. Working in ideal conditions (e.g., calm water, suitable viscosity), the recovery rate is between 42 and 66 gallons per minute, depending on whether a Class V or Class VII MARCO skimmer is used.

ODI SKIMMER Towed by two vessels, the ODI skimmer is a boom system using a barrier to contain oil for recovery. It features built-in weirs connected to discharge hoses enabling the oil to be recovered by a pumping system.

OWOCRS Open Water Oil Containment And Recovery System
Owned and operated by the Coast Guard, OWOCRS is a rapidly deployed, very rigid high seas containment barrier that also can be used as a skimmer. OWOCRS moves in a "U" configuration, towed by vessels at each end. A pump float subsystem is attached to the bend of the "U" The subsystem pumps oil collected by skimming weirs inside the "U" at a rate of 825 gallons per minute into a suitable container, usually a tank barge or a towed bladder. OWOCRS can be made stationary by means of a mooring system.

SEA PAC BARRIER SYSTEM   A Sea Pat Barrier System is a 23-foot boat containing 1,475 feet of inflatable boom that is automatically deployed and towed at three to four knots per hour

TK-5 PUMPS   The TK-5 pump used by the Coast Guard and the ADAPTS system share some of the same components. The TK-S pump functions as a corrosive liquid transfer system that also pumps viscous oils and high-temperature fluids.

VIKOMA SEA SKIMMERS   Vikoma sea skimmers utilize a rotary disc system to attract oil. Discs rotate oil toward a centralized pumping system which then pumps it into a suitable container. In optimal conditions, the best recovery rates for this unit is 220 gallons per minute.

VIKOVAK   A VIKOVAK is a high-performance vacuum for shoreline cleanup.

WALOSEP WEIR SKIMMER   Walosep Weir Skimmers use gravity to drain oil from the water surface. Using a pump to draw water-separated oil into the system, this skimmer has an oil recovery rate of up to 264 gallons per minute.

APPENDIX D. DISPERSANTS

Dispersants are chemical solutions used to reduce the cohesiveness of oil slicks. They are designed to remove oil from the surface of the water. Oil treated with a dispersant enters the water column as fine droplets which are then-distributed in three dimensions and subjected to natural processes such as biodegradation.

When the application of dispersants is technically feasible, its selection as a response option follows the choice between leaving the spill untreated and floating on the surface of the water where it may threaten surface resources, or treated and distributed in the water column where it may threaten subsurface resources. The threat posed by the dispersed oil droplets to subsurface resources in the water column is moderated by the relatively low toxicity of the present generation of dispersants and the dilution factor.

Dispersant use has been controversial since its notable introduction as a response option following the sinking of the tanker *Torrey Canyon* off the United Kingdom in 1967. Various studies since then have shown that the dispersants used in this incident caused more harm than good because of their indiscriminate use and/or highly toxic characteristics. Improved dispersants now in use are less toxic and more effective.

This response option generally has not been used in the United States for oil spill responses because of logistical difficulty, complex decision making or uncertain weather conditions-or because trial applications and evaluation indicate a lack of effectiveness.

When an evolving response strategy seems to indicate an advantageous use of dispersants, decisions must be based on rigorous technical, biological, and administrative considerations. The On-Scene Coordinator (OSC) initiates the decision process for dispersant use with a 'recommendation for concurrence' to the Environmental Protection Agency (EPA) and state Regional Response Team (RRT) representatives. The Department of Commerce and Department of the Interior, as natural resource trustees, may also be consulted.

To get RRT concurrence, a balance of many variables must be considered. It must be concluded that harm from dispersant use on the subsurface environment is likely to be less than the potential harm of untreated oil. Other factors that must be considered are spill location, type, and volume; time elapsed since the spill; existing and predicted wea-

ther; water temperature; salinity and sedimentation. The RRT also will take into account tides and tidal currents; risks to biological, physical, and economic resources; surface and subsurface trajectories: and availability of dispersants, application equipment, and trained personnel.

When they consider an OSC's 'recommendation for concurrence' to use dispersants, federal officials follow a decision making process established in the National Oil and Hazardous Substances Pollution Contingency Plan (NCP, 40 CFR 300, Subpart H). The NCP contains a general authorization-for-use policy and an EPA listing of dispersant's (the "Product Schedule') that may be authorized for use on oil spills.

The NCP also encourages each RRT to develop regional guidelines for preauthorized use of dispersants-allowing the OSC to forego RRT concurrence under certain circumstances, because it had already been obtained in the planning process. Criteria for preauthorization may be a combination of geographic, seasonal, biological, or other factors. The intent of preplanning and preauthorization is to minimize the interval between the time of the oil spill and the time of dispersant application because dispersant effectiveness may decrease dramatically relative to the elapsed time.

Preauthorized dispersant use decision making to reduce response time was pioneered by the Alaska RRT. The RRT had developed a comprehensive Regional Contingency Plan (RCP) dispersant use annex ("Dispersant Use Guidelines") with input from industry and fishermen's associations. The Guidelines divide Prince William Sound into three zones, each with a different preauthorization process (see Figure 10).

Zone 1 - The use of dispersants is acceptable and is preauthorized by the RRT at the discretion of the OSC

Zone 2 - The use of dispersants is conditional to protect sensitive resources. Dispersant use is likely to cause less harm in this zone than would result from its non-use. Prior to authorization, the OSC is required to submit a formal proposal to obtain the concurrence of the EPA and Alaska Department of Environmental Conservation (ADEC) RRT representatives.

Zone 3 - The use of dispersants is not recommended. Dispersant use is likely to cause more harm in this zone than would result from its non-use. Prior to authorization, the OSC is required to submit a formal proposal to obtain the concurrence of the EPA and ADEC RRT representatives.

The evaluation of an incident-specific dispersant application is a qualitative and subjective undertaking. The retrospective analysis of both decision making and operational effectiveness should avoid comparisons with earlier case histories which either cross a generation gap on the dispersant development timeline or involve a different set of incident-specific variables. Each RRT post-incident analysis should evaluate the dispersant section of the OSC report required for major oil spills, draw conclusions, and offer recommendations for national, regional, and local improvements.

National dispersant use policy guidance for coastal OSCs and RRTs was distributed by the Coast Guard in October 1987 in a "white paper" entitled *Dispersant Use Considerations* (Commandant (G-MER), unpublished). The text recognizes the importance of regional autonomy in the development of comprehensive dispersant use contingency planning at the regional and local levels and is intended to provide a measure of consistency to the approach used by RRTs.

The Interagency Technical Committee for the former EPA Oil and Hazardous Materials Simulated Test Tank (OHMSETT), whose membership includes the Minerals Management Service, the Coast Guard; EPA, the U.S. Navy, and Environment Canada, commissioned a National Research Council study entitled *Using Oil Spill Dispersants on the Sea.* The study was published in February 1989 and addresses the ecological, esthetic, and economic elements of dispersant use in open water. It also includes an assessment of the adequacy of dispersant-application technologies available for spill response. The study draws conclusions and offers recommendations on all aspects of dispersant use. The study is currently under consideration by the Interagency Technical Committee.

References:

Fraser, John P. "Methods for Making Dispersant Use Decisions," *Proceedings of the 1989 Oil Spill Conference,* American Petroleum Institute, Washington, DC., 1989, pp. 321-330.

Manen, Carol-Ann, (et al.). "Oil Dispersant Guidelines: Alaska," *Oil Dispersants -New Ecological Approaches,* ASTM STP 1018, L. Michael Flaherty, Ed., American Society for Testing and Materials, Philadelphia, 1989, pp. 144-151.

National Research *Council (NRC). Using Oil Spill Dispersants on the Sea.* National Academy Press, Washington, DC., 1989, 335 pp.

U.S. Coast Guard (USCG), Commandant (G-MER). *Dispersant Use Considerations.* Washington, DC., 1987, (Unpublished, 18 pp.).

APPENDIX E. STATE OF ALASKA
DISPERSANT USE DECISION MATRIX

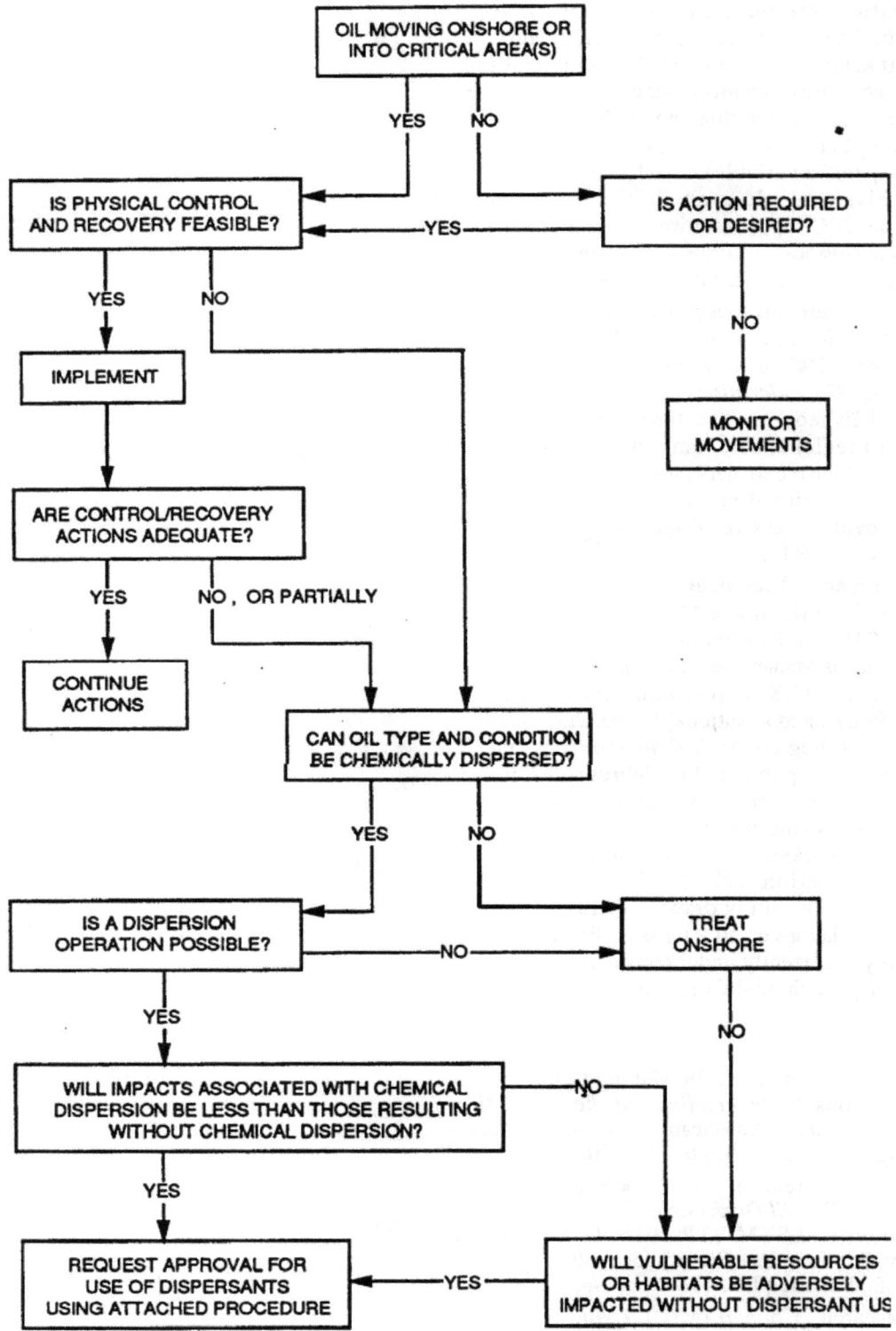

SOURCE: ALASKA REGIONAL CONTINGENCY PLAN

APPENDIX F. NATIONAL RESPONSE TEAM MEMBERS

Department of Agriculture	Mr. Bill Opfer
Department of Commerce (NOAA)	Mr. George Kinter
Department of Defense	Mr. Brian Higgins
Department of Energy	Mr. Richard Dailey
Environmental Protection Agency	Mr. Jim Makris, Chair
Federal Emergency Management Agency	Mr. Craig Wingo
Department of Health and Human Services (ATSDR)	Ms. Georgi Jones
Department of the Interior	Ms. Cecil Hoffmann
Department of Justice	Ms. Sheila Jones
Department of Labor (OSHA)	Mr. Frank Frodyma
Department of State	Mr. Bob Blumberg
Department of Transportation (Coast Guard)	Capt. Richard Larrabee, Z ngi1G lenv
Department of Transportation (Research and Special Programs Administration)	Mr. Alan Roberts
Nuclear Regulatory Commission	Mr. Bernard Weiss

APPENDIX G. ACRONYM GLOSSARY

ACOE:	Army Corps of Engineers
ADAMHA:	Alcohol, Drug Abuse, and Mental Health Administration
ADAPTS:	Air Deployable Anti-Pollution Transfer System
ADDS:	Air Deliverable Dispersant System
ADEC:	Alaska Department of Environmental Conservation
AIREY E:	See Appendix C, Equipment Glossary
ANILCA:	Alaska National Interest Lands Conservation Act
ANS:	Alaskan North Slope
AOSC:	Alyeska Oil Spill Coordinator
ATSDR:	Agency for Toxic Substances and Disease Registry
CDC:	Centers for Disease Control
CERCLA:	Comprehensive Environmental Response, Compensation, and Liability Act (Superfund)
CG:	U.S. Coast Guard
COTP:	Captain of the Port
CWA:	Clean Water Act
DOD:	Department of Defense
DOE:	Department of Energy
DOI:	Department of the Interior
DOMS:	Director of Military Support
EPA:	Environmental Protection Agency
FAA:	Federal Aviation Administration
FDA:	Food and Drug Administration
FOSC:	Federal On-Scene Coordinator
FWS	Fish and Wildlife Service
FY:	Fiscal Year
IBRRC:	International Bird Rescue Research Center
IMO:	International Maritime Organization
JTF:	Joint Task Force
LANTAREA:	Atlantic Area (USCG)
MAC:	Military Airlift Command.
MARCO:	See Appendix C, Equipment Glossary
MSO:	Marine Safety Office (USCG)
M/V:	Motor Vessel
NCP:	National Oil and Hazardous Substances Pollution Contingency Plan (National Contingency Plan)
NOAA:	National Oceanic and Atmospheric Administration
NRC:	National Response Center
NRC:	National Research Council
NRS:	National Response System
NRT:	National Response Team
NTSB:	National Transportation Safety Board
OSC:	On-Scene Coordinator
OSHA:	Occupational Safety and Health Administration
OWOCRS:	Open Water Oil Containment and Recovery System
PACAREA:	Pacific Area (USCG)
RCP:	Regional Contingency Plan
RRT:	Regional Response Team
SSC:	Scientific Support Coordinator
SUPSALV:	Supervisor of Salvage (U.S. Navy)
TAPAA:	Trans-Alaska Pipeline Authorization Act
TAPS	Trans-Alaska Pipeline System
T/V:	Tank Vessel
USCG:	U.S. Coast Guard
VIKOVAK:	See Appendix C, Equipment Glossary
VTS	Vessel Traffic Service

APPENDIX H. SHORELINE IMPACTS

Exposed Rocky Shores

o Common along open coastal areas of Prince
   William Sound and the Gulf of Alaska.
o Composed of steeply-dipping to vertical bedrock.
o Exposed to moderate to high wave action.
o Barnacles, mussels, snails, and various species of
   algae are common in most areas.

Predicted Oil Impact

o Most commonly, oil will be held offshore by
   waves reflecting off the steep cliffs.
o On less steep shores, oil may come onshore.
o Oil persistence is related to the incoming wave
   energy; during high-wave conditions, oil persistence *is* limited to days.
o Oil trapped in tidal pools will kill resident organisms.
o The damage to the intertidal community is expected to be relatively light, with fairly rapid
   recovery.
o Diving birds using these rocky sites may be killed
   if oiled.

Recommended Response Activity

o On most shores, no cleanup is necessary (and may
   be dangerous).
o Access is usually difficult.
o Cleanup of recreational areas may be necessary;
   high-pressure water spraying is effective while
   oil is still fresh.

Exposed Wave-Cut Platforms

o Very common along exposed portions of inner.
   Prince William Sound and the Gulf of Alaska.
o Consist of wave-cut or low-lying bedrock.
o May be very wide due to large tidal range.
o Commonly contain   narrow,   mixed-sediment
   beaches along the high-tide swash zone.
o The lower intertidal zone contains extensive algal
   growth.
o Tide pools and organisms are common in the
   lower-to-middle intertidal zone.

Predicted Oil Impact

o incoming oil commonly will form a band along
   the high-tide swash line.
o Tide-pool organisms may be killed.
o Lower intertidal algae may escape damage, depending on tidal stage and oil type and quantity.
o Oil persistence is limited (days to weeks) in most
   high-energy areas.

Recommended Response Activity

o In most wave-exposed areas, cleanup is not necessary.
o Removal of organisms should be avoided.

Source: Adapted from NOAA, 1983. Sensitivity of Central Environments and Wildlife to Spilled Oil, Prince William Sound, Alaska.

Mixed Sand and Gravel Beaches

o Very common throughout the study site.

o Present in both sheltered and exposed areas.

o Common as a narrow beach or stringer to top of bedrock platforms.

o Composed of coarse-grained sand, gravels of varying sizes, and possibly shell fragments.

o On active beaches, organisms are scarce due to the harshness of the environment.

o In stable habitats, algae may be attached to the larger gravel or boulder components.

o The larger rocks also may provide habitat for mussels, crabs, and snails.

Predicted Oil Impact

o Oil will be deposited primarily along the high-tide swash zone.

o Under very heavy accumulations, oil may spread across the entire beach face.

o Oil percolation into the beach may be up to 60 cm in well-sorted material.

o Burial may be very deep along the berm.

o Biota present may be killed by the oil, either by smothering or by lethal concentrations in the water column.

Recommended Response Activity

o Remove oil primarily from the upper swash lines.

o Removal of sediment should be limited.

o Mechanical reworking of the sediment into the wave zone and/or high pressure water spraying can remove the oil effectively; sorbent boom may be necessary to capture oil outflow.

Gravel Beaches

o Fairly common throughout areas dominated by bedrock.

o Composed of gravel of varying sizes.

o Shell fragments and woody debris also are common beach components.

o Biomass generally is very low in high-wave areas; at calmer sites, the population of fauna and attached algae may be fairly great: crabs, snails, mussels, barnacles, and attached algae are most common.

Predicted Oil Impact

o Under light-to-moderate oil concentrations, oil would be deposited primarily along the high-tide swash zone.

o With heavy oil quantities, the entire beach face may be covered.

o Oil may percolate rapidly and deeply (up to 1 m) into the beach face.

o If oil is left to harden, an asphalt/gravel pavement may result.

o Resident fauna and flora may be killed by the oil.

Recommended Response Activity

o Removal of sediment should be restricted.

o Pushing gravel into the active surf zone and use of high-pressure water spraying is effective in removing oil while it is still fresh.

o Sorbent boom should be used to capture oil outflowing during the cleansing process described above.

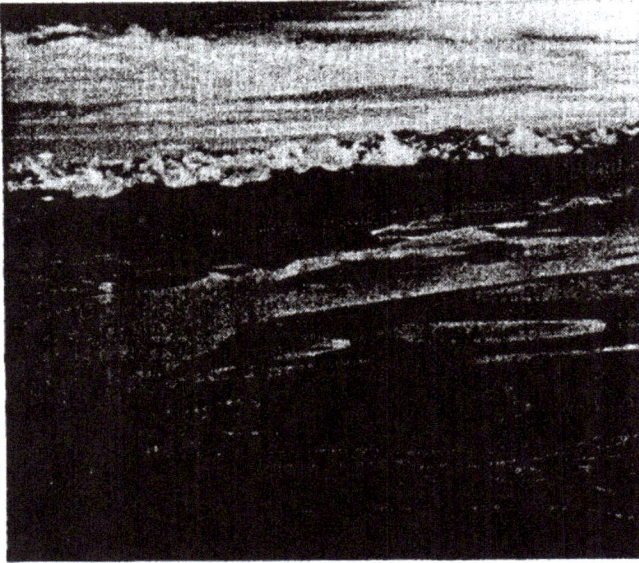

Exposed Tidal Flats

o Particularly common in Orca Inlet and in front of the Copper River delta.

o Also found on the very wide uplifted bedrock platform fronting Kayak Island.

o Visible only at low tide.

o Exposed to low-to-moderate wave energy and/or tidal currents.

o Composition is most commonly sand or mixed sand and gravel.

o Species density and diversity may be high; soft-shelled clams and worms are most important.

o Many millions of migrating birds use these flats as a seasonal feeding ground.

Predicted Oil Impact

o Most oil will be pushed across the flat as the tide rises.

o Deposition of oil on the flat may occur on a falling tide if oil concentrations are heavy.

o Biological damage may be severe.

Recommended Response Activity

o Cleanup is very difficult (and possible only during low tides).

o The use of heavy machinery should be restricted to prevent mixing oil into the sediments.

o On sand flats, oil will be removed naturally from the flat and deposited on the adjacent beaches where cleanup is more feasible. In gravelly areas, oil may bind with the sediment; high-pressure water spraying may be necessary.

Sheltered Rocky Shores

o Composed of bedrock outcrops, ledges, or boulders.

o Common within the sheltered, interior portions of Prince William Sound.

o Species density and diversity vary greatly, but barnacles, mussels, crabs, snails, and rockweed often are very abundant.

Predicted Oil Impact

o Oil will persist for several years, especially between rocks.

o Upper intertidal biota and algae will be the most severely affected.

o Algae present in the lower intertidal zone are most resistant to damage.

Recommended Response Activity

o These areas need priority protection using deflection booms, sorbent booms, and offshore skimmers.

o High- and low-pressure water spraying is effective while oil is still fresh.

o Cutting of oiled algae is generally not recommended.

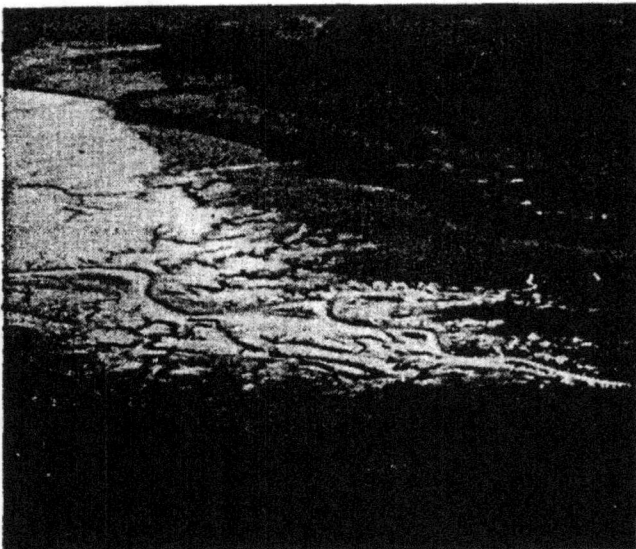

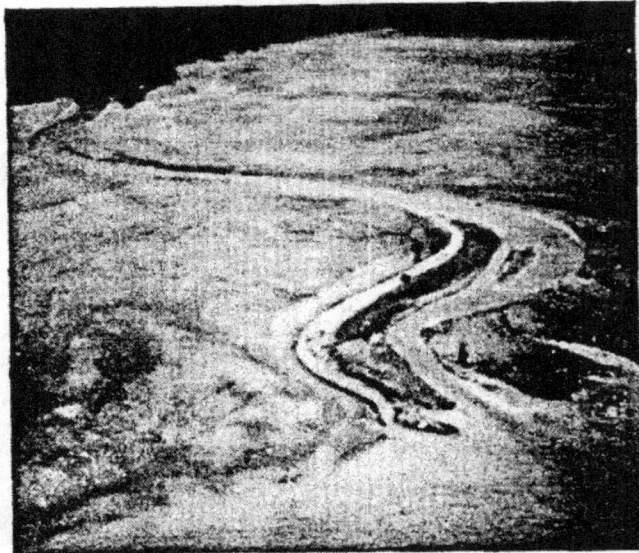

Sheltered Tidal Flats

o Very common in the upper portions of Orca Inlet, adjacent to marshes of the Copper River delta, and in the upper portions of many fjords.
o Present in calm water habitats, sheltered from major wave activity.
o Composed of muds.
o Usually contain large populations of razor clams, worms, and snails; commercial harvesting -of shell fish occurs on many of these flats.
o Seasonally, bird life is very abundant (millions) in the Copper River delta/Orca inlet area.

Predicted Oil Impact

o Oil may persist for many years.
o Incorporation of oil into tidal-flat sediments over the long-term is common.
o Oil deposition commonly will occur along the upper fringes of the flat.
o Very heavy oil accumulations will cover much of the surface of the flat.
o Biological damage may be severe.

Recommended Response Activity

o This is a high-priority area necessitating the use of spill protection devices to prevent or limit oil spill impact; open-water deflection, sorbent booms, and open-water skimmers should be used.
o Cleanup of the flat surface after oiling is very difficult because of the soft substrate.
o Manual operations from shallow draft boats may be helpful.

Marshes

o Small marshes common at the head of many fjords; broad fringing marsh along the Copper River delta.
o Very sheltered from wave and tidal activity.
o Composed primarily of Spartina grasses on an rich organic mud base.
o The Copper River delta region supports over 20 million migratory shorebirds and several hundred thousand waterfowl.
o Marshes are nursery grounds for numerous fish species; also, crabs are particularly common.
o The Copper River delta area is the only nesting area of the dusky Canada goose.

Predicted Oil Impact

o Oil in heavy accumulations may persist for decades.
o Small quantities of oil primarily will be deposited along the outer marsh fringe or along the upper wrack (debris) swash line.
o Resident biota, including bird life, is likely to be oiled and possibly killed.

Recommended Response Activity

o Under light oiling, the best practice is to let the marsh recover naturally.
o Cutting of oiled grasses and low-pressure water spraying is effective, especially during the early part of the spring growing season.
o Heavy oil accumulations on the marsh surface should be removed manually; access across the marsh should be greatly restricted.
o Clean-up activities should be supervised carefully to avoid excessive damage to the marsh.

www.ingramcontent.com/pod-product-compliance
Lightning Source LLC
Chambersburg PA
CBHW081340090426

42737CB00017B/3232